SpringerBriefs in Applied Sciences and Technology

SpringerBriefs present concise summaries of cutting-edge research and practical applications across a wide spectrum of fields. Featuring compact volumes of 50 to 125 pages, the series covers a range of content from professional to academic.

Typical publications can be:

- A timely report of state-of-the art methods
- An introduction to or a manual for the application of mathematical or computer techniques
- A bridge between new research results, as published in journal articles
- A snapshot of a hot or emerging topic
- An in-depth case study
- A presentation of core concepts that students must understand in order to make independent contributions

SpringerBriefs are characterized by fast, global electronic dissemination, standard publishing contracts, standardized manuscript preparation and formatting guidelines, and expedited production schedules.

On the one hand, **SpringerBriefs in Applied Sciences and Technology** are devoted to the publication of fundamentals and applications within the different classical engineering disciplines as well as in interdisciplinary fields that recently emerged between these areas. On the other hand, as the boundary separating fundamental research and applied technology is more and more dissolving, this series is particularly open to trans-disciplinary topics between fundamental science and engineering.

Indexed by EI-Compendex, SCOPUS and Springerlink.

Benjamin Steyn

Technology vs Nature

The Third Political Spectrum

Springer

Benjamin Steyn
School of Advanced Study
Institute of Philosophy
University of London
London, UK

ISSN 2191-530X ISSN 2191-5318 (electronic)
SpringerBriefs in Applied Sciences and Technology
ISBN 978-3-032-18581-5 ISBN 978-3-032-18582-2 (eBook)
https://doi.org/10.1007/978-3-032-18582-2

This Springer imprint is published by the registered company Springer Nature Switzerland AG
The registered company address is: Gewerbestrasse 11, 6330 Cham, Switzerland

Praise for *Technology vs Nature*

"There can be few subjects as important in our rapidly changing contemporary world as technology and nature. And the two are often seen to be in conflict. But why do we value them? Benjamin Steyn argues that a key ingredient is missing from our understanding and that this hampers our ability to shape policy; do we regard technology and nature as intrinsically or merely instrumentally significant? This book is a vital contribution to that debate."

—David Edmonds, *Oxford University, Author of Would You Kill the Fat Man? and Co-host of Philosophy Bites*

Acknowledgements I am indebted to academic colleagues with whom I've discussed the content of this book over many years. Particularly Kristoffer Ahlstrom-Vij, Barry Maydom, Rob Simpson, Fabienne Peter, Robert Northcott, John Basl, Reid Hoffman, Matthew Bradley, Shahar Avin, Alex Grzankowski, Emma Borg, Nick Shea, and Barry Smith. I am grateful for editors, reviewers, and publishers at the *Journal of Philosophy and Technology* and the *Journal of Environmental Ethics* who advanced the papers upon which this book is built. And to Anthony Doyle and colleagues at Springer for everything in preparation for this book. I am grateful to colleagues at the Department for Science, Innovation and Technology, UK Research and Innovation, Department for Environment, Agriculture and Rural Affairs, and Department for Energy and Net Zero, who, over the years, have inspired this work with their real-life passion, frustration, and political acumen on the subjects of nature and technology. Thanks most of all to my wife, family, and friends for tolerating years of me talking about this.

Competing Interests The author has no competing interests to declare that are relevant to the content of this manuscript.

About This Book

This book proposes that 21st-century political debates in AI, bioethics, and environmental policy are best understood as occurring along a new political spectrum: a zero-sum conflict between the intrinsic value of nature and technological progress. The political values spectrum is a simple conceptual tool that everyone already knows. However, many modern political debates are not adequately explained by traditional economic or social liberty spectra. Instead, they are better illuminated by this new spectrum, concerning technological mastery over the biology of our bodies, minds, functions, and physical environment. By defining the metaphysical opposition of nature and technology, and building an original case for technology's intrinsic value as a form of physical liberty and collective human achievement, this broad work seeks to unify disparate ethical fields, provide clarity in political discourse, and show that the persistent struggles in philosophy of technology, environmental ethics, and bioethics to draw a principled line between acceptable and unacceptable technological interventions is in fact a political problem of balancing competing intrinsic goods, not a failure of ethical reasoning.

Contents

About the Author

Dr. Benjamin Steyn has a dual career as a UK government official and academic. He currently leads the UK Metascience Unit at the Department for Science, Innovation and Technology, where he funds and conducts "research-on-research" aiming to accelerate scientific progress for social good. He has previously held senior jobs in the Department for Environment, Food and Rural Affairs, Department for Business, Energy and Industrial Strategy, and Department for Exiting the EU, and was a founding member of the UK Advanced Research and Invention Agency. Academically, Benjamin is currently a Visiting Post-Doctoral Fellow at the Institute of Philosophy, School of Advanced Study, University of London, where he is interested in technological and environmental ethics, philosophy of science, philosophy of AI, and political science. He holds a Ph.D. in philosophy and political science; master's degrees in environmental economics and philosophy; and an undergraduate in economics and political science.

Part I
Constructing the Nature Technology Political Spectrum

Chapter 1
Political, Intrinsic, Value Spectrums

Abstract This chapter addresses a significant casualty of disciplinary siloing across political philosophy, political science, and analytical ethics: the lack of a rigorous, critical examination of the political values spectrum as a conceptual tool. Starting from a purely descriptive, non-normative premise, it advocates for the political spectrum not merely as a pragmatic reductionist model for political scientists but as a powerful apparatus for unifying ethics and political theory. It achieves this by precisely defining the metaphysical character of political values. The core argument advanced is that political values are a unique subset of intrinsic ethical values, structurally defined by their existence as antagonistic pairs that constitute a zero-sum trade-off. This logical incompatibility, which prevents the simultaneous maximization of both poles (e.g., economic liberty versus equality), fundamentally distinguishes political values from ordinary intrinsic goods (e.g., culture versus healthcare). To prevent the unjustified expansion of the standard two-dimensional model (economic and social liberty), the chapter concludes by establishing four "blind judging criteria" that any candidate spectrum—including the proposed nature-technology spectrum—must satisfy to demonstrate genuine explanatory power and timelessness within political discourse.

Keywords Political philosophy · Political theory · Axiology · Intrinsic value · Political spectrum · Political values · Zero-sum trade-off · Metaphysics of value · Descriptive political science

I have a dual career as a UK government official and academic philosopher. Over the last decade, the two have converged. Both in terms of subject focus—scientific and technological progress and the politics thereof, and also in terms of my approach to both kinds of work.

I lead the UK Metascience Unit, an internal government research lab and funder jointly run by the Department for Science, Innovation and Technology, and UK Research and Innovation. The field of metascience applies scientific methods to the study of the research system itself to improve policy. Here, having my own "lived experience" of being an academic has been a great help.

B. Steyn, *Technology vs Nature*,
SpringerBriefs in Applied Sciences and Technology,
https://doi.org/10.1007/978-3-032-18582-2_1

But more importantly, my philosophical life is shaped by my government life in some unique ways:

First, as a civil servant, I have been politically neutered. I'm literally not allowed to express a first-order normative political position, for fear of breaking the impartiality conditions of the UK Civil Service Code. This makes me uniquely suited to describing and clarifying battlefields rather than doing battle. This is a divergence from most political philosophy, which is normative and not particularly descriptive.

Second, while I don't have normative political freedom, I have unusual intellectual freedom. My work in metascience has shown me that academia is incredibly siloed. There is low-hanging fruit to be found between disciplinary boundaries, but the incentives to publish and be cited in discipline-specific journals suppress this type of work. Being only part-time in a university, I am fortunate to be less bound by these incentives. I can take the risky move of writing at the intersections. That is what I do in this book, and if I succeed, I would take no offence to one describing it as grabbing the low-hanging fruit. I think the argument is so simple and obvious, but it takes an unusually interdisciplinary view to get there.

Third, I am committed to the idea that political philosophy should endeavor to be practically useful in the hands of the policymaker. And this means placing a premium on two features: accessibility and empirical testability.

Policymakers can engage readily with the language of social science. Many central government officials are social scientists. Even those who aren't will often do similar tasks: apply frameworks to reduce complexity to fundamental concepts and relationships; collect and analyze empirical data; and, in some cases, perform experimental methods to evaluate interventions. Both share a principle that such evaluation should strive for sober and dispassionate objectivity.

Political science aims to reduce the complexity of political thinking into descriptive models by isolating the core variables that drive political behavior. In recent decades, political scientists have advanced that the fundamental predictors of political attitudes are a set of core political values. The field has generally coalesced around a reduction of these values into two spectra: one of economic liberty (economic left/right) and the other of social liberty (authoritarian/libertarian).

These value spectra are intuitive and intelligible in day-to-day political life, and that is a great virtue of them as a thinking tool. In characterizing stances as being, say, "left wing" or "libertarian", those in the political sphere can mentally map politicians, parties, and the public with relative ease. Political spectra give a common language. But not only are they a mere convenience, they are also scientifically robust. Across decades and contexts, they yield internally consistent results, showing that correlations across multiple survey questions are driven by these single underlying constructs.

The reduction of political ideology to political spectra puts political scientists and laypeople at odds with traditional political philosophers, for whom the landscape of ideas is too rich for simple models. Thinkers like Mill, Marx, Berlin, Rawls, Nozick, and Sandel cannot be neatly decomposed or precisely mapped onto a minimal set of dimensions. Moreover, scientific reductionism is absolute anathema to continental philosophers, who tend toward "world-building" and "concept positing", requiring lifetimes to understand, let alone empirically test, commensurate, compare, and contrast within a common framework. This esoteric complexity is why political philosophy is generally not read by those engaged in political life and has minimal impact as a result.

We can see political science and political philosophy as two academic silos talking past one another. To complicate matters, a third silo, analytical ethics and axiology (the study of values), is also disconnected to the latter two. When political scientists and philosophers discuss values, they fail to speak precisely about their metaphysical character or their relationship to ethical values. The unfortunate casualty of this disciplinary siloing is that no single field has the tools or motivation to examine the most intuitive piece of conceptual apparatus used in actual political life: the political values spectrum.

This book is about nature and technology, but it is also a love letter to the political spectrum. The humble political spectrum is not just a convenient reduction for political scientists, parliamentarians, and civil servants, but a powerful conceptual tool that unifies ethics with political philosophy and in turn, political science, and from there, a route into political reality.

To understand the hidden power of the political spectrum, we must first address the question rarely asked within each silo: what are political values?

This starts with axiology, the philosophical study of value. Axiology provides the classic distinction, introduced by G.E. Moore, between intrinsic value and instrumental value. At its core, intrinsic value is value held for something "for its own sake", or "as an end in itself", rather than as a means to some further end (instrumental value). Philosophers use the term to denote that the valued thing is objectively valuable in a grander ethical sense; its existence improves the world.

When something has intrinsic value, someone has a good reason to try to promote it, or preserve it, or make it true, or bring it into existence. Adding something with intrinsic value to the world makes the world or a life or an outcome better (other things equal). Something's having intrinsic value does not mean this value always trumps other considerations, but it gives reasons *pro tanto*. Intrinsic values are additive, admit of degrees and can be positive or negative, and could enter a hypothetical "value calculus". Many sorts of things have been argued by philosophers to be of intrinsic value: utility, knowledge, beauty, life or humans or persons, love, culture, equality, friendship, etc.

For a classical utilitarian, all these *prima facie* intrinsic goods are ultimately only instrumentally valuable toward the single intrinsically valuable property: utility. This commensurability allows values to be easily traded off, aligning with our intuition that compromise is necessary. But many philosophers are value-pluralists, citing distinct and incommensurable realms of value, capturing the rivaling intuition that we would not sacrifice Y (say, moral integrity) for any amount of X (say, aesthetic pleasure).

What about *political* values—economic equality, social liberty, community? It is clear they are intrinsic values. Proponents want them for their own sake and are willing to sacrifice other things for them. The crucial question, which falls between the many silos, is this: What distinguishes a political value (like social liberty) from a wider set of intrinsic values (like life, love, beauty, or truth)?

One might define political values as those concerned with co-operation and conflict within societies. But this is insufficient. Societies exercise power and channel resources over values like truth and beauty. Perhaps they are values over which consensus is necessary? Again, this is too generous; all ethical values can be sources of collective disagreement.

Should we simply accept that the class of political values is a large, ill-defined subset of all ethical values? No. I propose that the unique, distinguishing feature of political values is that they are structurally related in antagonistic pairs. Each political value occupies one end of a zero-sum trade-off with another. These pairs pit two *prima facie* desirable things against each other, creating an inherent and endless source of tension that defines the political realm.

Consider the spectrum of libertarianism and authoritarianism. If somebody values personal freedoms from the state (negative liberty), this is logically incompatible with a prescriptive authoritarian state. An increase in negative liberty comes, by definition, at the expense of state control. Power is either centralized or decentralized. Similarly, conferring economic liberty directly opposes state management designed to maintain equality; as Nozick famously declares, "liberty upsets patterns" [1]. Even with infinite political power and resources, it is logically impossible to maximize both outcomes simultaneously.

This is not the case for ordinary ethical intrinsic values. While intrinsically valuable things like "culture" and "healthcare" compete for resources, if there were an infinite budget, there would be no contradiction in pursuing both maximally. The conflict in political values is not just scarcity; it is a logical necessity that prevents both ends of the spectrum from being achieved simultaneously. Political values are political values because they have an opponent.

Thus, we are starting to see that political spectra are more than just pragmatic reductions; they are powerful conceptual tools that bridge the gap between ethics and political philosophy. Political values are Moorean intrinsic values, but they are also values that exist in pairwise metaphysical trade-offs. They are not, like other intrinsic values, states of affairs to be maximized. Rather, they describe states of affairs that, for the vast majority of us, must be *balanced*.

The economic liberty and social liberty spectra have gained prominence because economics and social policy are eternal areas where societies must agree to cooperate. However, new spectra may reflect new defining battlegrounds as political debate finds new vectors and consensus forms in others. Perhaps multiculturalism versus assimilation or hawkish versus dovish foreign policy could be candidate 21st-century spectra. While these things may be only instrumental for some, they may also be underpinned by principles that constitute a genuine spectrum.

For the virtues of accessibility and simplicity, we must be careful not to expand the simple two-by-two model unnecessarily. We should only posit new spectra if they truly contribute to better political understanding and debate, and conceptual clarity among competing ideological proponents.

To avoid an unjustified expansion, here are four blind judging criteria that any candidate spectrum must meet. Over the course of this book, we will be exploring whether my proposed spectrum, pitting nature against technology, meets these criteria.

1.1 Blind Judging Criteria for a Candidate Political Spectrum

a. Genuine Trade-Off

The primary feature is the zero-sum trade-off. While all goods compete for resources, the opposition on a political spectrum must be mutual incompatibility, where one side's gain is roughly equal to the other's loss, driven by logical necessity, not mere scarcity.

b. Significant Additional Explanatory Power

The new spectrum must be orthogonal to incumbent political spectra (e.g., the classic 2×2 economic/social liberty model). There must be political questions whose stances are *not* satisfactorily predicted by current dimensions but *are* by the new one. This explanatory power can be tested theoretically, by cleaving apart different interests or, quantitatively, by revealing that the new dimension explains a significant proportion of variance in attitudes, as revealed, for instance, through voting behavior. We fail if, for instance, we show the spectrum reducing neatly to existing left–right axes.

c. The Possibility of Intermediate Positions

For the debate to be represented as a spectrum rather than a binary choice, it must be able to accommodate and differentiate a range of views between the extremes. This allows for the possibility of moderate views, for instance, a population being normally distributed along such a spectrum. But it could be that we see polarization, i.e., a double-humped distribution.

d. Timelessness

Political spectra must describe a stable and generalizable feature of political discourse, not something reducible to specific spatiotemporal political decisions (e.g., stances on High Speed Rail 2). While real-world issues are multi-faceted, stances must be justified by reference to the stable spectrum, reflecting eternal conflict.

With these criteria set, and the political (intrinsic) value spectrum in hand, we can now move to the actual subject matter of this book: nature and technology and the politics and value thereof.

Reference

1. R. Nozick, *Anarchy, State, and Utopia* (Basic Books, 1974)

Chapter 2
The Politics of Nature and Technology

Abstract This chapter introduces the central idea of this book: the Nature and Technology Intrinsic Value Spectrum. It begins by identifying a convergence of 21st-century political questions—spanning debates on Artificial General Intelligence, human enhancement, geoengineering, environmental policy, de-extinction, and synthetic biology—which share no common thread within conventional political discourse. It argues that while these issues contain layers of economic, social, and human welfare considerations, they are fundamentally united by visceral, principled reactions concerning the intrinsic value (or disvalue) placed on the natural world and on technological endeavor. The chapter formalizes this observation into a core claim: the existence of a new political spectrum orthogonal to the traditional economic and social liberty axes. This spectrum features an intrinsic value for Nature (or Naturalness) at one end and an intrinsic value for Technology (or Technological Endeavor) at the other, proposing that conflicts between these intrinsic values provide strong, distinct reasons for action. This structural tension is a timeless feature of political disagreement, tracing its historical roots to the Enlightenment's ethos of mastery and the Romantic critique of natural boundaries. Through contemporary examples from public policy and ethical debate—including marine conservation, the "unnaturalness critique" of geoengineering, the divergence in sentiment toward humanoid robotics, and the pro-human bias in appraising AI art—the chapter substantiates the necessity of this new model for a comprehensive, descriptive political understanding.

Keywords Intrinsic value · Philosophy of technology · Environmental ethics · Political spectrum · Nature-technology divide · Bioethics · Geoengineering · Political axiology · Technological progress · Naturalness critique

There is a new jumble of political questions rearing their heads in the twenty-first century. Each has its own lively discourse, but no common thread has been articulated. This jumble of questions is around the value we place on the natural world, and on humanity's technological endeavors. They are about the limits that ought or ought not to be placed on human control of nature, and the preservation of natural spaces, objects, and processes. They are questions that have always been part of political

B. Steyn, *Technology vs Nature*,
SpringerBriefs in Applied Sciences and Technology,
https://doi.org/10.1007/978-3-032-18582-2_2

discourse to some extent but have risen in importance now as technological advancement gives us more and more potential to alter fundamental biological processes at a macroscopic and microscopic level and substitute human activity with artificial intelligence. They are questions like, but by no means limited to, the following:

Is it important that humanity protects animals, plants, and their natural habitats? Is it important to conserve wildernesses like forests, jungles, and coral reefs? Do we advocate genetic modification of crops? Should the solution to climate change be predominantly technological, for instance, using geoengineering to alter the weather artificially, or come from humans reducing their impact on the world? It is important for taxpayers to fund blue-skies feats of innovation and engineering, and prioritize this over spending the same money to improve basic public services like health, education, and defence? Should we strive for Artificial General intelligence? Should we tweak our own brain chemistry for purposes other than medicine? Should we be able to choose the color of our babies' eyes and hair, and screen it for genetic defects? Should we be able to change our biological sex? Is it inherently worse living in a virtual reality simulation? Genetic engineering will soon allow us to de-extinct a woolly mammoth using DNA frozen during the ice age. Should we bring back these creatures? …and what if the same question was posed for de-extincting our humanoid relative, the Neanderthal? And what of cloning human beings? Or growing them outside a womb?

Each of these questions will be pregnant with many considerations of ethical value, such as for the value of "human welfare." For instance, many consider the existential risks posed by the rapid acceleration of artificial intelligence; the importance of nature conservation to human health; and the risks to human welfare posed by not innovating to solve climate change. Each question will also bear in some way on political values. What will be the distributional economic impact of allowing human enhancement technologies to enter a free market? What will be the impact on individual liberty if states wield the power of AI? What does social media do for the integrity of a community?

But all these questions are also united by one common thread of value consideration. In all cases, these questions evoke visceral, emotional, principled reactions (both positive and negative) to do with the activity's status as a natural or technological affair. It is these visceral reactions that I will advance that are telling of a value (or disvalue) for these matters in their own right—a valuation separate from human welfare—narrowly defined—and separate from the political values already mentioned.

Nature and technology are political matters, because they concern the use of public resources, legislation, governance, and strategy. However, as a society, we are relatively unpracticed in our treatment of these issues as overtly political. And it takes an interdisciplinary view, one traversing ethics, political theory, and political science, to advance that these nature and technology debates are better conceived as political, rather than ethical.

2.1 Nature and Technology Intrinsic Value Spectrum

My core claim in this book is that a political spectrum exists that explains a significant proportion of the variance in attitudes to many nature and technology questions. This spectrum features, at one end, a high intrinsic value placed on nature (or naturalness), and at the opposing end, a high intrinsic value placed on technology (or technological endeavor).

Justifying that this is indeed a single spectrum, I have to show that there is a genuine conflict here. I also have to show that this spectrum is orthogonal (i.e., logically independent) to our other well-trodden political spectra, economic left/right, and social authoritarian/liberal. That is, we must be able to prove that all eight octants of this political value space must be possible and coherent. And this three-dimensional graph should be sufficiently more useful to navigate politics than our usual two-dimensional political model, covering just axes of economic and social liberty (Fig. 2.1).

Recall, the axiological underpinning of political spectra is captured in the ethical concept of intrinsic value. Intrinsic value is the value placed on some object or state of affairs as an end in itself. This should ward off a knee-jerk response to the proposed spectrum: it is undeniable that an *instrumental* value for either technology or nature is not at odds with either instrumental or intrinsic value for the other. Naturalists will most likely be strong advocates of technologies like solar panels or wind farms because they reduce the pace of climate change and so protect their desiderata: nature.

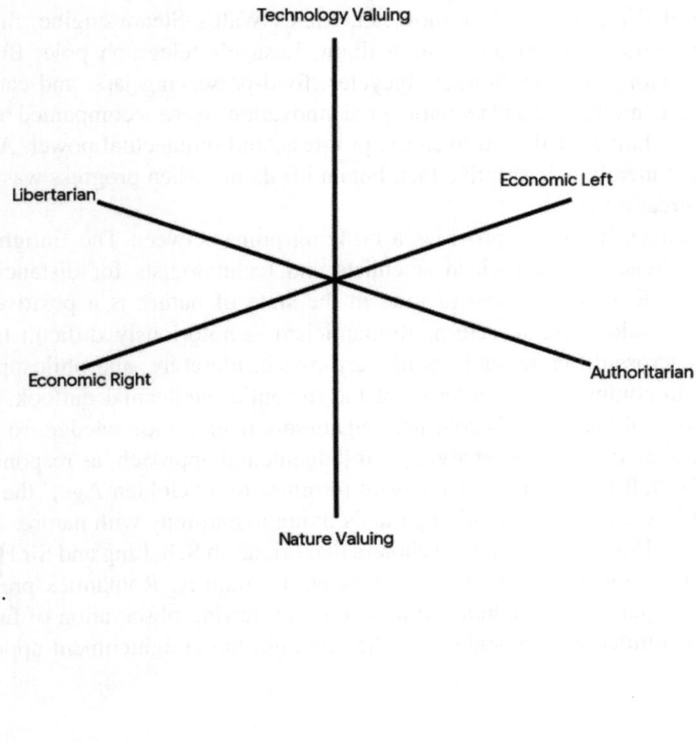

Fig. 2.1 .

Technologists will place some value on nature, for the innumerable instrumental services it provides us with. But again, this is valuing nature instrumentally.

In other cases, the two may find some common ground—solar panels might be valued by technologists as an instance of a successful well-designed technological pursuit, and by naturalists for saving natural processes from climate change. Nature and technology debates resemble other political debates in being multi-faceted, and any given policy in the real world might be advocated by both sides for different reasons, in pursuit of different instrumental and intrinsic values. The complex web of instrumental and intrinsic values involved in nature and technology debates can never be fully reduced to a single spectrum. However, I will advance that conflicts between intrinsic attitudes to nature and technology do provide strong reasons for action in a variety of cases and will matter even more in a volatile age of environmental destruction and rampant technological progress.

While the concept of nature and technology as opposing political values may seem intuitively simple, its formal articulation as an orthogonal political spectrum is novel. This novelty stems partly from the unique context outlined in Chap. 1. But it also may stem partly from the tendency of historians to view this tension as a time-bound phenomenon happening between the 18th Century Enlightenment and Romanticism, rather than a timeless, structural feature of political disagreement.

The Enlightenment, also called "The Age of Reason," was a period of unprecedented intellectual achievement, including the work of Francis Bacon, Isaac Newton, Leonhard Euler, Rene Descartes, Denis Diderot, John Locke, Adam Smith, David Hume, Thomas Reid, and Immanuel Kant. It was an age where we also saw great technological achievement, including Benjamin Franklin's bifocal glasses and lightning rod, Gabriel Farenheight's Thermometer, James Watt's Steam engine, the Montgolfier Brothers' first hot air balloon flight, Lesage's telegraph pole, Bushnell's submarine, along with guillotines, bicycles, food-preserving jars, and carbonated drinks. These intellectual and technological innovations were accompanied by a positive belief in human self-assertiveness, progress, and intellectual power. Advancement wasn't merely a descriptive fact, but an ideal, and when progress was made, a matter of great pride.

Jean-Jacques Rousseau provides a bridging point between The Enlightenment and Romanticism. He criticized scientists and technologists for distancing man from nature. Rousseau's view of man in the state of nature is a positive one to which we should strive to return. Romanticism is notoriously difficult to define and spans across different subjects like art, music, literature, and philosophy with very little in common. But, in terms of the romantic intellectual outlook, Romantics cite The Enlightenment's cold-hearted attempt to extort knowledge from nature and its rational, calculating, analytic, and judgemental approach, as responsible for humanity's "fall from grace." They want to return to a "Golden Age," the biblical Garden of Eden, a time symbolizing man's living in harmony with nature.

Following Rousseau, romantic scholars like Friedrich Schelling and Sir Humphry Davy held an uneasy relationship with scientific inquiry. Romantics preferred a "hands-off" approach to understanding nature, preferring observation of facts with careful and limited experimentation. They thought the enlightenment approach to science a problematic attempt to control nature. Understanding nature required an attitude of admiration, love and worship; always to observe, never to control.

Mary Shelley, daughter of enlightenment philosopher William Godwin and feminist Mary Wollstonecraft and wife of romantic poet Percy Shelly, wrote Frankenstein [5] at 19 from an upbringing immersed in the culture clash. The character Dr. Victor Frankenstein is a derogatory portrayal of The Enlightenment. Dr. Frankenstein is vehement in the pursuit of knowledge and power to create life. He is a megalomaniac, wanting to play God by breaking natural boundaries that ought not to be broken.

The mistake we have made is to see these Enlightenment and Romantic movements as historically rooted. There is in fact nothing temporally or spatially anchored. Fast forward to the present day, where our advances in human enhancement biotechnology make Frankenstein's monster a reality, participants in bioethics debates use the same sentiments without hyperbole—a subject to which we will return later in the book.

Turning now to the present, nature and technology are both areas of public policy that, as a UK Government official, I have had the pleasure of working on over the years. On numerous occasions, and in different roles, I have seen how people can talk past each other due to a failure to distinguish between their principled and practical concerns for each. While working in the International Marine Team on marine biodiversity conservation and ocean plastic pollution, I found a split in opinions among stakeholders. There were those for whom marine nature ought to be treated purely as a commonly owned economic resource, as typified in the popular jargon "ecosystem services." Coral reefs have a quantifiable monetary value in providing tourism, flood defences, carbon sequestration, and in providing breeding grounds for fish, which, in turn, are economically valuable as food.

For others, however, the relationship with the ocean is better described as a deep love and fascination. I found many times in government where it is an open question whether or not we should apply economic evaluation criteria to a decision, such as whether to invest UK Aid budget to save coral reefs in Indonesia. For the ocean lovers, the notion of reducing such decisions to economics was often either abhorrent, absurd, or a drastic undervaluation. Indigenous communities in the south pacific worship the ocean as a deity—we can't conveniently collapse such appreciation to a monetary value. Sometimes such tensions would rise to the surface for a ministerial decision. It is not uncommon for an environment minister to see beyond the reduction of nature to an ecosystem service, but day-to-day actions of officials will sometimes unwittingly and unintentionally rule on this ideological question in particular circumstances, either one way or the other.

More recently, while working in the Department for Science, Innovation and Technology, I have seen a divergence of views similar in structure among some stakeholders. There are those I meet who are principled advocates of technological endeavor, for no other reason than because it is an exciting quest for humanity to be on. They want to invest public money in space travel, artificial enhancement, quantum computing, and all sorts regardless of the clear public benefit, narrowly defined. Humanoid robotics is a paradigmatic case: there seem to be some economic benefits (personable household assistants for the elderly, for instance), but compared to other emerging technologies, they seem to garner a disproportionate sense of wonder and fascination among researchers, policymakers, and, in some cases, the public. Others, meanwhile, are more narrowly interested in these technological endeavors only to the extent that they create economic prosperity or serve some other need such as national defence, and others still have strong reservations about the thought of technological advance.

In the case of both nature and technology policymaking then, it is not clearly articulated whether we are undertaking the endeavors in question for the sake of some ultimately economic or social end, or whether, in addition to this, these activities are more fundamentally worth pursuing…for their *own* sake. My government colleagues working on land management policy face similar daily arguments over levels of urbanization of rural landscapes, in decisions over the building of new towns, cities, factories, quarries, shopping centers, and transport links across natural (or at least more natural) habitats like woodlands, plains, or mountain ranges. The differences between urban and rural life are manifold and will bear on many of the items on our hypothetical list of intrinsic values. But the defining feature of urbanity is immersion in technology. When we decide to pave over a woodland to make a new urban development, we face these questions of value.

Elsewhere in environmental policy, we can also see the nature/technology tension in debates over climate change mitigation. Historically, naturalness has been a recurrent theme in debates around nuclear energy. Social scientists commonly surveyed public opinion about the health hazards of natural and man-made radioactivity, finding numerous examples where the public are disposed against unnatural radioactivity, for instance, the radioactivity in nuclear facilities, even when radiation levels are lower than in natural instances of radioactivity, for instance, in soils and fossil fuels.

More recently, the "unnaturalness critique" has been explicit in debates around geoengineering—activities such as routinely pumping a chemical mist into the sky to simulate the climate-cooling effects of a volcanic eruption. Some offer a techno-optimistic vision of geoengineering. Meanwhile, others object not only because it is practically risky, but also in principle.

Even when the resistance to geoengineering is not explicitly labeled as principled pro-nature/anti-technology stance, it can be seen implicitly. There is a sense among anti-geoengineering writers that there is something morally deficient about a technological solution to climate change. A commonly cited idea is around "green moral hazard": If geoengineering solutions seem feasible, this allows a generation to "evade their moral obligations" to stop climate change by cutting carbon emissions. The term "techno-fix" is used in some cases pejoratively to imply a sticking plaster to the climate problem and not a "genuine solution." But what manifests as a principled objection to a "techno-fix" might be considered a failure of imagination that in fact a technology can, quite unproblematically and reliably, solve a problem. We used to have a moral obligation to drive slowly before we invented seatbelts, and we used to have a moral obligation to limit the amount of sex we have with our partners before we invented contraception. What motivates those who seek to moralize the use of techno-fixes? As we will see, it's a deep-rooted value for nature.

Beyond environmental policy, the same sentiments can also be seen as a timeless feature of decisions around new technological developments. The term "Luddite," originally referred to a group of 18th-century workers who destroyed industrial machinery. But it now exists as a derogatory term for those opposed to technological development. However, we now have a loose movement termed neo-luddism, reclaiming the word, in rejecting technology and embracing simple natural living. This includes Chellis Glendinning, pioneering the concept of "ecopsychology," who is critical of the inherent negative psychological effects of technology in *My name is Chellis & I'm in recovery from Western civilization* [1]. At the more extreme end, such sentiments have been voiced by Unabomber American terrorist Ted Kaczynski, who submitted to the Washington Post a 35,000-word manifesto endorsing the r*ejection of all modern technology; Rejection of civilization itself; Rejection of materialism, and Love and reverence toward nature or even worship of nature* [2]. Though it may be easy to dismiss the extreme conclusions expressed here, the neo-luddites are correct to make explicit that the corollary of rejecting technology is embracing nature. While they are *prima facie* an anti-technology movement, their message is identical to those we would consider firmly in the camp of pro-nature philosophers.

In direct contrast to neo-luddites, we can consider those who clearly see technological progress as a virtue. Consider these quotes from famous engineers:

> The engineer has been, and is, a maker of history.
>
> —James Kip Finch, American engineer and educator
>
> Strive for perfection in everything you do. Take the best that exists and make it better. When it does not exist, design it.
>
> —Sir Henry Royce, English engineer and car designer
>
> We are continually faced by great opportunities brilliantly disguised as insoluble problems.
>
> —Lee Iacocca, American engineer and automobile executive.
>
> One man's "magic" is another man's engineering. "Supernatural" is a null word.—Robert A. Heinlein

These quotes seem to display an intrinsic appreciation of the pushing of technological boundaries, not in pursuit of some particular end state, but for the act of creation itself.

The expansion of our capabilities in biotechnology is an area where the nature and technology sentiments particularly come alive, because biological entities are things that, ordinarily and historically, have been natural entities, and their naturalness is being threatened (or positively revolutionized) in sudden leaps and bounds by technological advancements. Later, we will be taking a deep dive into debates around human enhancement biotechnology, where there has been a high quantity of philosophical discussion around whether the activity's unnaturalness ought to be a reason not to do it.

But for now, consider some other applications of biotechnology. We now are nearing the technological capability to be able to de-extinct species by cloning and artificially growing their DNA. There are also interesting and potent questions about whether or not we ought to actively create new artificial life. For some scientists and policymakers in the field that I have spoken to in my work, the creation of artificial life is extremely interesting and exciting. Others baulk at the very idea, and when we drill down, a common reason offered is a fear arising from the activity's unnaturalness.

Also consider ethical debates around the medicalization of pregnancy. Some advocate "natural" childbirth, free from technological assistance and doctors. Others argue that these groups fail to explain why natural birthing is better or more ethical owing to the fact it is more natural. Even if one rejects the presumption in favor of natural birthing, they may not extend themselves to endorsing the other extreme, which is the emerging technology of ectogenesis: an artificial womb outside the body. In an interview with Vice magazine, ectogenesis philosopher Anna Smajdor notes that she received hate mail from multiple different political groups one would expect to disagree on most things. She remarks that *"when you challenge or threaten this very, very deep-seated idea about a woman's role, about the beauty and value of natural childbirth and motherhood, you stir up some extremely passionate and aggressive responses in people"* [4]. I think the fact that these objections cut across standard political demographics and archetypes is telling.

Lastly, consider a recent similar divergence in public sentiments on AI, for instance, on the topic of AI art. Millet et al. [3] surveyed the public, asking for their appraisal of pieces of AI and human art. They find a pro-human (anti-AI bias), where no matter which piece of art is in fact human made, they prefer the art that is labeled as human made. They suggest that people tend to think of creativity as an intrinsically valuable, and intrinsically natural human, faculty. The authors, meanwhile, see this as a prejudice that ought to be eliminated, referring to the human bias in a derogatory fashion, following environmental ethicists, as anthropocentrism.

These featured examples are only scratching the surface. We can also consider everyday divergences in sentiments around genetically modified crops versus the organic food movement, cosmetic surgery versus a natural beauty ideal, and on issues of the weight given to naturalness in debates on gender identity. AI, given its generality, is already spawning a vast multitude of examples, so much so that new political language is urgently required to engage constructively in these debates.

So, the mission should be clear: This book will build toward the conclusion that these nature and technology debates are best explained with reference to a political values spectrum.

The remainder of Part 1 will arm readers with the key concepts: we've now had the idea of intrinsic value—borrowed from ethics and axiology, and the idea of the political value spectrum—borrowed from political science. Next, we will consider definitions of nature and technology borrowed from their respective philosophies. Once these building blocks have been clearly stated, we will then be able to see the interplay between them.

In Part 2, we will consider in what sense can either nature or technology be intrinsically valuable? We will make sense of decades of rich (but often confusing) literature on the intrinsic value of nature, and then we will provide some original arguments to support the technology side, which is comparatively underappreciated.

Once the theory is firmly on the table, Part 3 will show the spectrum's immense explanatory power. How the spectrum view supersedes, resolves, and reduces some inferior distinctions and concepts in environmental ethics and bioethics, such as the empty and misleading concept of "anthropocentrism." We take a deep dive worked example of the spectrum's utility in diagnosing a *prima facie* complex ethical debate, which is, *should we embrace human enhancement biotechnologies?*

This is followed by a discussion comparing and contrasting the claims of this book against other key thinkers working at the intersection between nature and technology. In the realm of anglophone analytical philosophy, there is much environmental ethics and some normative philosophy of technology, but little uniting them both. In the realm of continental philosophy, there is more work addressing the two, however these thinkers are not intending to offer precise descriptive claims that can be operationalized in political science. As we move from the theoretical into the empirical realm of political science, we also look at the inadequacy of current public attitude surveys in identifying the spectrum.

We conclude by seeing how this new spectrum measures up against the blind judging criteria set out above.

References

1. C. Glendinning, in *My Name is Chellis & I'm in Recovery from Western Civilization* (Shambhala, 1994)
2. T. Kaczynski, in *Unabomber Special Report: Industrial Society and its Future* (The Washington Post, 1997). https://www.washingtonpost.com/wp-srv/national/longterm/unabomber/manifesto.text.htm
3. K. Millet, F. Buehler, G. Du, M.D. Kokkoris, Defending humankind: Anthropocentric bias in the appreciation of AI art. Comput. Human Behav. 143(Article 107707) (2023). https://doi.org/10.1016/j.chb.2023.107707
4. N. Preskey, in *Why are Natural Births Still so Popular?* VICE Online. Retrieved 12 Mar 2022, from https://www.vice.com/en/article/439mvm/this-is-why-the-natural-birth-movement-isnt-slowing-down. Accessed 01 Dec 2025
5. M. Shelley, in *Frankenstein: Or, The Modern Prometheus.* Lackington, Hughes, Harding, Mavor, & Jones (1818)

Chapter 3
Defining Nature and Technology

Abstract This chapter establishes the core conceptual building blocks for the Nature-Technology spectrum by providing rigorous, reductivist definitions of nature and technology. First, nature is defined by its independence from human activity ("absence of human intervention"), which is defended as a continuum rather than a binary distinction. The chapter resolves the conceptual overlap between "naturalness"" and "wild(ness)" by advocating for their simplification into a single category. Second, technology is formally defined as a Tripartite Analysis: Intentional Functional Artifact. The chapter concludes by noting the structural analogy between this definition of technology and the classical Platonic definition of knowledge. Just like in the knowledge case, the tripartite definition of technology suffers from what we can call "techno-Gettier cases."

Keywords Environmental philosophy · Philosophy of technology · Definition of nature · Definition of technology · Naturalness continuum · Restoration thesis · Intentional functional artifact · Gettier problem · Epistemology of technology

Having assembled the first building block, the idea of a political value spectrum, we now bring in the next two: our concepts of nature and technology, what these terms mean and how they relate to one another.

3.1 Defining Nature

Consider some things that we typically might want to consider natural. A forest. A hippopotamus. A geological formation on a distant planet. A person's talent for running. What unites these things? The naturalness of some entity depends on its history and origin, in particular, how it came into being and what modifications it has gone through. What is "natural" broadly hinges on independence from human activity. For my examples: The forest is natural as it is land uncultivated by humans. The hippopotamus is an animal with a clear evolutionary lineage involving no human

B. Steyn, *Technology vs Nature*,
SpringerBriefs in Applied Sciences and Technology,
https://doi.org/10.1007/978-3-032-18582-2_3

selective breeding. Geological formation forms without human intervention. That someone's running ability can be described as a "talent" implies the person has not tried to actively cultivate it, but rather it has been endowed by nature. Naturalness in this sense is to be contrasted with "artificial": something that is the result of human activity. Contrast the examples above with a shopping center; a domestic dog; a hole in the ozone layer; the clearly acquired ability to tie a shoe.

Philosophers disagree which human activities will qualify as interfering with an entity's naturalness/artifactualness. We might begin with a stringent condition that naturalness is a "total independence from the intervention of human beings." But various philosophers point out that anthropogenic pollution and climate change, which have now affected all corners of the globe, mean that every place on earth is now in some sense human modified. Williams [19] argues that there is a paradox in even delineating certain areas as "nature," e.g., national parks, as even the act of protection is a human act. A strict "no humanity" approach would then be too exclusive to be useful.

Noting the stringency but wanting to maintain a binary distinction between naturalness and non-naturalness, many thinkers have given variants of a slightly weaker claim, that naturalness is independence from certain *types* of human activities [14]. If only a certain subset of human activities affects naturalness, what is this subset? Attempting to answer this question is where we glide from the mere descriptive attempt to define nature, to the normative attempt to define it in a way that is consistent with our intuitions about precisely what *ought to* be preserved. The attempt to define the concept is bound together with the valuing of it.

On matters of animal captivity, Weston [18] gives a nice example: Environmentalist David Brower, in rejecting the conservation of condors in captivity, said *"Condors are 5% flesh and bones and 95% place."* For Brower, extricating the condor from its habitat made it to a great extent unnatural, and for this reason, a misguided attempt at nature conservation. Here we can see two axes of potential disagreement. One might agree or disagree that nature is intrinsically valuable, but also, among those who do Weston's intuitions about natural intrinsic value, one could still contend that in fact "condors are 51% flesh and bones and 49% place," concluding on balance that it is beneficial to preserve them in captivity than extinct them.

In Elliot's seminal work "Faking Nature" [2], he offers a scenario where we are deciding whether to allow a company to mine a woodland, which involves destroying it, on the condition that they regrow the wildlife after. He begins by positing his strawman, the Restoration Thesis, which assumes that this mining is permissible, so long as the wildlife re-grown is of the same amount and type as before. Elliot thinks the Restoration Thesis is insufficient, likening the attempt to brand this restoration as "nature" to a forgery. A forged painting, while sharing exactly the same physical characteristics as the original, is less valuable. We not only care about the physical characteristics, but also the provenance, history, and authenticity of the painting. The restoration thesis is tantamount to forgery, says Elliot, in faking the proper processes that ought to have produced the woodland, namely, natural ones.

Others have followed suit. Lee [10] argues that natural entities (both organisms and inert matter) are more morally considerable than artifacts because they possess the "ontological value of independence," while artifacts are created by humans with explicit intention to serve their own interests and ends. Meanwhile, Rolston [12, 13] argues that what is valuable about natural entities is their having human-independent evolutionary processes.

Are these debates all-or-nothing affairs? Does the condor in captivity lose *all* its naturalness (and hence natural intrinsic value)? Does the regeneration project have *any* naturalness (and hence natural intrinsic value)? Or are the regenerated landscape and the captive condor only as valuable as, say, a carpark? The answers will depend on whether naturalness/artifactualness is viewed as a binary distinction, or on a continuum.

Elliot himself is one of many subscribing to a continuum view of naturalness: that something can be more or less natural depending on the amount of human activity involved in an entity's history. He thus concedes that the regeneration project holds *some* naturalness and hence value [2, 3]. Eric Katz also values naturalness intrinsically. But unlike Elliot, he argues that it is more an all-or-nothing matter (1998). Human attempts to mimic the natural world inevitably exhibit a human technological domination over nature, in a way that is in fact offensive to nature, and so we cannot produce anything of real naturalness ourselves.

Following what appears to be a critical mass of recent environmental philosophers, I have endorsed the continuum view and reject Katz's across my work on this topic. It seems intuitive to think that selective breeding practices, which have been practised for hundreds of years, are less natural than wild animal breeding, but more natural than artificial insemination, which, in turn, is more natural than lab-grown meat, which, in turn, is more natural than a future lab-grown super-protein. The fact that a given one of these methods of meat production might have a measure of human artifice doesn't prevent us from valuing its residual naturalness.

In a 2025 article in the Journal of Philosophy and Technology, I set out a short skeleton argument which became the precursor to this book [15]. In response to that article, Fernández-Jimeno and Domínguez [4] pointed out that there are other uses of the term "nature" which don't conflict with the artificial. They raise the example of "re-wilding" which is both "natural" and artificial. In my reply, I clarified that a casual use of the word "natural" to mean "biological/ecological" should be abandoned in technical discourse now that biotechnology has reached a new level of maturity [16].

Naturalness, properly defined, hinges upon some "absence of human intervention." But there is another term which crops up often in environmental ethics: "Wild(ness)" and "wilderness." Puzzlingly, there is clearly huge conceptual overlap between wild(ness)/wilderness and nature. And yet, many in the field seem to juggle the two terms concurrently and without acknowledgement of the crossover:

For instance, Hettinger and Throop [6] define "wildness" as the extent to which it is not "*influenced, altered or controlled by humans." They elaborate that "While one person walking through the woods does little to diminish its wildness, leaving garbage, culling deer, or clear cutting do diminish wildness, although in different degrees.*" They note in a footnote that: *"Although a number of philosophers have appealed to wildness and the related notion of naturalness, there is no uniform agreement on its meaning or justification."*

It is odd that this casual observation of conceptual overlap was made with no effort exerted to rectify it. They then proceed to give multiple arguments for the intrinsic value of wildness, all of which can be incredibly sensibly substituted for naturalness. As is clear from the quote above, they even suggest that wildness can admit of degrees, making it a perfect substitute into Elliott's nature/artifact continuum discussed above.

Meanwhile, Elliot, whose project revolves around defining and valuing naturalness, also often uses wilderness without unpacking the term. For instance, in *Faking Nature,* he says: *The appeal that many find in areas of wilderness, in natural forests and wild rivers depends very much on the naturalness of such places* [2, p. 86].

If you accept my claim that these terms are basically synonyms, this translates as: *The appeal that many find in areas [untouched by humans], in forests [untouched by humans] and rivers [untouched by humans], depends very much on [the extent to which such places are untouched by humans].* That is just to say the simple statement: "nature is appealing." In later work, Elliot [3] repeatedly uses the phrase "*Wild Nature.*" This phrase seems either tautological, or it is a clarification that we are talking about natural objects and not, say, human nature, or the nature of X as in the essential properties of X. But here, in the domain of environmental ethics, this seems redundant.

"Wild*er*ness" might be permissible as a distinct term denoting an extremely natural area, or to denote a barren and uninhabited landscape, for instance, a desert or a sterile distant planet. But this is not the sense that Elliot and Hettinger and Throop use. These philosophers are concerned with environmental ethics in the world at large, and they endorse the continuum view of nature/artifact. Their concern extends to residual naturalness all the way along the spectrum.

Beyond being a general inconvenience in this area, the failure to observe synonymity—or at least a very close conceptual connection—between "wild" and "natural" means we have a parallel discourse that causes confusion. In the spirit of Occam's Razor and scientific reduction, I prefer to simplify matters by placing all views that intrinsically value wildness/wilderness within a single category. This adds to our category new philosophical company, as many philosophers have argued for the moral significance of wilderness (see a range of views in an anthology *The Wilderness Debate Rages On* [11].

So, in summary, we have a working definition of nature as "absence of human intervention." This is something that admits of degrees and is metaphysically opposed to "artificial." And this nature is something which, as we will soon see, lots of people think is valuable.

3.2 Defining Technology

"Technology" has two distinct senses: abiding by the -ology suffix, it denotes a subcategory of knowledge, that field of study related to manufacturing and engineering. But "technology" is also more commonly and colloquially used as a descriptor of a certain class of physical objects (phones, cars, laptops, etc.). For thinkers focused on defining the latter sense, there is some consensus that the following concepts constitute technological objects [1, 7, 9, 17]:

(1) They are artifacts—human-made objects. As above, this is often recognized by scholars on both sides to be on a continuum with "nature"—i.e., that which is absent of human intervention [2, 14]. Something can be semi-artificial, and by extension, semi-technological. Like prepared food, a cultivated garden, or a domesticated pet.
(2) They serve a function. An object is used "for something," "serves a means to an end," or has some purpose.
(3) They were *intended* to serve the function. Micro-plastic pollution in the ocean is a human creation, but not an intentional one. It is then an artifact, but not technology. Similarly, junk, space debris, etc.

I am by no means the first to articulate these features, but the simplistic reductionist that I am, I want to propose that we can do away with any verbose description here and actually formalize this into a simple tripartite analysis: *Technology = Intentional Functional Artifact.*

We can see in the following table how this tripartite definition gets a variety of cases right:

	Intuitively technology?	Intentional?	Functional?	Artifact?	Match between intuitions and definition?
The Apollo 13 spacecraft	Yes	Yes, it was designed with the intention of flying astronauts to the moon	Yes, it successfully achieves the purpose of transporting astronauts	Yes, a spaceship is a man-made object	✓
A human heart	No	No	Yes, it has a biological function of pumping blood around a body	No	✓
Plastic pollution from microbeads	No	No, it is an unintended waste/ byproduct	No, it serves no commonly accepted function	Yes	✓

There is something that epistemologists might find exciting to notice about this tripartite formulation of technology. Those less interested in epistemology are welcome to skip ahead to Chap. 4 at this point.

The tripartite definition of technology is very similar in structure to the classical Platonic definition of knowledge as "justified true belief." We have in each case:

(1) A metaphysical object: a belief, an artifact.
(2) A property held by a subset of 1, specifically a subset that we value in some way. *true* beliefs; *Functional* artifacts.
(3) A process involved which reliably generates 1 + 2 and so rules out mere instances of 1 + 2 brought about by sheer luck. A good *justification* for believing something true; the right *intentions* in producing functional artifacts.

Maybe this structural similarity to a "decent" definition of knowledge should give us comfort that this definition of technology is somewhat robust. However, just like for "knowledge = JTB," this definition of technology similarly suffers from Gettier cases. Gettier cases [5] are instances of "lucky knowledge," owing to bizarre coincidences, where there are a complete severance of the belief's justification and the conditions that made it true.

Here are some examples of techno-Gettier cases:

Case 1: A watch facade (no inner workings) is intended to tell the time, fails to do so conventionally, but happens to be stuck exactly on the user's two specific check-in times (7:00 a.m. and 7:00 p.m.). The object is an Intentional Functional Artifact (Technology) but is intuitively junk.

Case 2: Bob manufactures a copper doorstop, intending it to be robust. The object is too light and smooth to work ordinarily. However, a user, Rachel, places it in an electric doorway. The doorstop is successful, not because it physically holds the door, but because the copper accidentally short-circuits the door's electronics, causing it to freeze exactly where Rachel wanted it.

In these cases we have an artifact (a watch/doorstop), intended to perform a function (to tell the time/to hold open a door), and succeeding in performing that function, and so, we have an intentional functional artifact, and hence technology. And yet, under normal circumstances, what has been produced clearly belongs on the scrapheap. It does not seem to be something we ought to value as an instance of technology.

Gettier cases have prompted myriad responses, but Linda Zagzebski's seminal paper [20] proves that Gettier cases are unavoidable for all JTB + accounts of knowledge. One can always generate them so long as there is some degree of independence between justification and truth, Gettier cases are always constructable. It is analogous in the case of techno-Gettier cases. So long as there is some degree of independence between the intentions of a maker/user of an artifact and its true functionality, we can always generate hypothetical cases of "double luck," whereby an object's functionality and the intentions of its maker/user counteract each other in precisely the right way, such that the artifact is rendered both functional and intended to be so, and yet remains counterintuitive as a successful technological endeavor. Taking heed of Zagzebski's warning, I do not suppose we can solve Gettier or techno-Gettier problems here, and in both cases I accept the tripartite definition as the best we currently have. Though I would speculate that if a solution to one came along, it would analogously solve the other.

While this note about techno-Gettier cases is intended to be a fun aside, the serious point here is that the conceptual analogy between technology and knowledge will become important to the argument later when we consider a technological version of the "meno problem": what makes knowledge (technology) more valuable than mere true belief (functional artifact).

References

1. R. R. Dipert, in *Artifacts, Art Works, and Agency* (Temple University Press, 1993)
2. R. Elliot, Faking nature. Inquiry: An Interdisciplinary J. Philosophy **25**(1), 81–93 (1982)
3. R. Elliot, Intrinsic value, environmental obligation, and naturalness. Monist **75**(2), 138–160 (1992)
4. N. Fernández-Jimeno, M.G. Domínguez, Which technologies are valuable and for whom? Value plurality and contingency in technological design. Philosophy Technol. **38**(2), 1–6 (2025)
5. E.L. Gettier, Is justified true belief knowledge? Analysis **23**(6), 121–123 (1963). https://doi.org/10.1093/analys/23.6.121
6. N. Hettinger, B. Throop, Refocusing ecocentrism. Environ. Ethics **21**, 3–21 (1999)
7. R. Hilpinen, Authors and artifacts. Proc. Aristotelian Soc. **93**, 155–178 (1993)
8. E. Katz, The big lie: Human restoration of nature. in *Nature as Subject: Human Obligation and Natural Community*. ed. by E. Katz (Rowman & Littlefield, 1997), pp. 93–107
9. P. A. Kroes, A. W. M. Meijers, (Eds.), The dual nature of technical artefacts [Special issue]. Stud. History Philosophy Sci. **37**, 1–158 (2006)
10. K. Lee, in *The Natural and the Artefactual: The Implications of Deep Science and Deep Technology for Environmental Philosophy* (Lexington Books, 1999)
11. P. N. Michael, J. B. Callicott, in *The Wilderness Debate Rages on: Continuing the New Great Wilderness Debate* (The University of Georgia Press, 2008)
12. H. Rolston, in *Philosophy Gone Wild: Essays in Environmental Ethics* (Prometheus, 1986)
13. H. Rolston, in *A New Environmental Ethics: The Next Millennium for Life on Earth* (Routledge, 2011). https://doi.org/10.4324/9780203804339
14. H. Siipi, Dimensions of naturalness. Ethics Environ. **13**(1), 71–103 (2008)
15. B. Steyn, The nature technology political spectrum. Philosophy and Technol. **38**(1), 1–25 (2025)
16. B. Steyn, Author's reply to 'which technologies are valuable and for whom? Value plurality and contingency in technological design' by Fernández-Jimeno & Domíguez, a commentary on the paper 'The Nature Technology Political Spectrum' by Benjamin Steyn. Philosophy and Technol. **38**(2), 1–4 (2025)
17. P.-P. Verbeek, P. E. Vermaas, Technological artifacts. in *A Companion to the Philosophy of Technology*, ed. by J. K. B. Olsen, S. A. Pedersen, V. F. Hendricks, (2009). https://doi.org/10.1002/9781444310795.ch28
18. A. Weston, Between means and ends. The Monist **75**(2), 236–249 (1992). www.jstor.org/stable/27903288
19. B. Williams, Must a concern for the environment be centred on human beings? in *Making Sense of Humanity and Other Philosophical Papers* (1995)
20. L. Zagzebski, The inescapability of Gettier cases. The Philosophical Quart. (1950-), **44**(174), 65–73 (1994)

Chapter 4
The Opposing Metaphysics of Nature and Technology

Abstract This chapter provides the metaphysical foundation for the Nature-Technology Political Spectrum by demonstrating that its two poles are in genuine zero-sum conflict for physical resources and processes. It begins by establishing that all physical matter exists on a continuum between nature (independence from human activity) and artifact (human-made objects). The core argument is that technology, as the predominant category of artifacts, fundamentally opposes nature by reducing the naturalness of the world through both the conversion of natural matter into artifice and the augmentation or replacement of natural processes. While acknowledging common ground against non-functional or unintended artifacts (e.g., pollution), the chapter concludes that the allocation of all residual physical material remains an antagonistic, zero-sum problem. Finally, the chapter tackles key metaphysical challenges, including the objection that artifice is intrinsic to human nature and the problem of locating the human being on the continuum. It resolves this by concluding humans are semi-natural and semi-artifactual, maintaining that choices regarding technology increase or decrease humanity's relative proportion of naturalness. This proven, orthogonal metaphysical trade-off between the two values is illustrated by the incompatibility of the Naturalist and Technologist utopias with each other, while each of these utopias is fully compatible with all four quadrants of the traditional political spectra.

Keywords Metaphysics of nature · Technology production · Nature-artifact continuum · Human nature and artifice · Political utopias

Having defined nature and technology, we should now be able to see that the two are in a deep metaphysical trade-off.

As we have seen already, it is common for philosophers to conceive all physical matter as located somewhere on a continuum between nature and artifact. Nature is that which is not the result of human endeavor/interaction. Artifacts are that which are human made. The organic lemon is more natural than the genetically modified lemon, which, in turn, is more natural than the lemon chemical food additive, but

B. Steyn, *Technology vs Nature*,
SpringerBriefs in Applied Sciences and Technology,
https://doi.org/10.1007/978-3-032-18582-2_4

none we might think are as natural as the lemons which grew for hundreds of years undiscovered in ancient Asia.

Artifacts oppose nature in both physical matter and processes. Artifice takes as inputs objects which can generally be described as less artificial than its outputs. So artifice reduces the amount of nature in the world. Given a state of affairs with a certain stock of physical material, each artifact reduces the naturalness of that state of affairs.

The less obvious but equally important case concerns intervention/non-intervention in the *processes* carried out by entities in the physical world. Many artifacts are functional. By adding a new function to the user's disposal, a function which was not already endowed upon users in nature, these functional artifacts alter, augment, circumvent, or replace natural processes or capabilities. Given a certain stock (or flow) of natural objects and processes, each artificial intervention in augmentation of, or replacement of, this stock reduces levels of naturalness.

Technology is the predominant sub-category of artifacts. *Ceteris paribus*, increasing the quantity of one reduces the quantity available to be classified as the other. We can see quite straightforwardly then that technology and nature are metaphysically opposed. But is this trade-off actually entirely zero-sum (i.e., one side's gain is equal to the other's loss)? There are some caveats to consider.

Notice that the nature-technology value spectrum does not entirely supervene on the nature artifact metaphysical continuum. There are categories of objects that are artificial, and either not intentional or not functional or both. An increase in unintended pollution worldwide is desirable for neither side. Anthropogenic climate change, if taken as an unintended artificial process, is desirable for neither as it looks likely to both destroy nature and destroy economies: which are needed to stimulate production of new technologies. Non-recyclable wastes, i.e., artifacts that no longer serve a function, are desirable for neither. There might be no appreciation from either side for a dead wasteland caused by an anthropogenic but accidental occurrence of acid rain.

There is thus much that can be agreed on between the two sides, to ensure that the maximum possible stock of physical material is dedicated either to nature or technology and not neither. That said, while they can agree to minimize unintentional and non-functional artifacts, the question still remains: how do we allocate any residual physical material between nature and technology? Again, the answer puts the two sides in zero-sum conflict.

There is a further kink. This account assumes that there are not some further valuable artifacts that fall outside the subset of "intentional" and "functional." On a broad definition, functional objects might include artistic and ornamental objects, but some will want to claim these objects are not technology but nonetheless valuable. This would threaten the one-dimensionality of the nature/technology value spectrum by creating a third valuable class that is competing for some stock of physical material.

My first line of response is that I am inclined to suggest art is functional and so does fall within technology. My second response is that this category is relatively insignificant compared with nature and technology, and for our purposes we can simply re-draw the battleground upon which nature and technology are fighting on, to exclude art and cover the politics of "*most* of our physical world."

This sketch of a metaphysical account faces yet more challenges. It is hard to determine either the relative or absolute position of a given object or process on this continuum. We might intuitively want to say that a rural countryside town is more natural than a concrete-clad city. Certainly, if we blended up the contents of each and analyzed the resultant mush, the countryside would come out as containing more *biomass* than the city. But the countryside and its biomass have been subject to a long history of cultivation by humans—longer than cities. So how can we say they are more natural? An account of the natural/artificial status needs to suggest how we weigh certain types of human manipulation of matter over time and space. That account is unfortunately not forthcoming.

Another objection to this continuum is that one might argue it is *part of* human nature to create artifacts. If true, this leads to contradiction, as then artifacts are in some sense natural. This idea is advanced by Jose Ortega y Gasset, who argues *"man without technology is not man"* [5, p. 96]. It is intrinsically human to react upon one's environment, not to resign himself to the world as it is. And a similar argument is made by Stiegler [8] who argues that technology is not an addition to human life, but the original condition of the human being itself.

These arguments should be partially resisted. It is certainly true that some amount of artifice is part of human nature: consider our opposable thumbs designed for toolmaking, or hunter-gatherers' firemaking leading to cooking—digestion outside the body. But to say that it is necessarily in our nature to engage in artifice *always,* begs the normative question at stake in this book. The extent to which we engage is an open question. We are at all times both engaging in it and resisting engaging in it. Gasset and Stiegler only consider the instances where we do engage. But we could give an equal and opposite narrative to explain how the resistance to technology and reverence for nature has come to exist. Why, for instance, has it taken so long in human history to fly? Current levels of artifact production (or the production of any given artifact) could not easily be said to be the result of an entirely *natural* process, but rather the relative triumph of one deep-seated human socio-psychological characteristic (to create) over another (to leave this as they are).

Turning the nature/artifact metaphysics onto humans themselves raises another set of potential objections. If all physical material sits on this continuum, this also of course must then include the physical stuff of human beings (and their constituent parts, their minds, offspring, excretions, etc.). If we cannot successfully locate human matter on the continuum, the metaphysics is broken. We might be initially inclined to say humans are natural. Humans have evolved through *natural selection*—a process that predates humans and so, by definition, cannot be an artificial one. The challenge for someone who rejects this claim is to state where along the evolutionary lineage artifice crept in? We could, for instance, consider the evolution of some property gained by human ancestors, (e.g., metacognition), as being where unnaturalness started. But even this looks like a gradually acquired faculty. For this reason, the idea that the human retains the property of naturalness is initially attractive.

However, evolution also involves many occurrences of one set of humans making or modifying another. Most obviously we make one another through reproduction, and thus successive generations of humans are the artifacts of those before them. There are various other ways in which humans modify and, both literally and figuratively, make each other. As Simone De-Beauvoir said: “One is not born, but made a woman” [3, p. 283]. Nurture has a profound physical effect on body and brain development. We also modify each other’s bodies through physical violence, applying social pressure to lose weight, or through giving each other haircuts, piercings, tattoos, amputations, and cosmetic surgery.

But humans cannot be purely artificial either. A pure artifact is an impossibility as anything we call an artifact in the real world must have at some point had natural inputs. Things we call artifacts day-to-day are things with a high degree of human involvement. Where we use the label artificial is typically a pragmatic linguistic choice, relative to context. If we declared humans to be artificial, this would hinder attempts to consider humans separately to obviously *more* artificial forms of intelligent agents, for instance, AI or cyborgs, that look soon to become a reality.

Let us say then that humans are semi-natural and semi-artifactual. We are products of nature and nurture. There are elements of our physical makeup and process that we undergo that are natural and have remained unchanged since the days of our pre-artifactual ancestors. But we are also to some extent artificial and proceed to literally make and modify each other’s physical instantiations through our conscious activity. We don’t necessarily need humans as a species to occupy a single fixed point on the spectrum. Successive generations of humans, and different instances within the species at a particular time, might have greater or lesser proportions of naturalness and artificiality.

This view avoids the problems of the extreme positions. By rejecting that humans are *purely* natural objects, the view avoids the contradiction of human reproduction as artifice. It also accepts that some human-on-human behavior is artifice, while accepting that some naturalness is preserved in humans, and remains unchanged. The position also retains the intuitive usage of terminology like human nature and artificial insemination.

Locating particular humans in an exact spot on the continuum is a fuzzy and fraught affair, even more so than it is for any other physical object. But all that is required for the forthcoming axiological discussion is not an absolute mapping, but a relative mapping. All we need to say about humans is that they are somewhere in the middle, and choices we make will pull them up or down, and in so doing, increase or decrease relative proportions of natural and technological value.

In Part 2 of this book, we will be diving into the normative arguments advocating for each side of this spectrum, breathing life into what has thus far been an abstract discussion of nature and artifact metaphysics. But before we do, let us just spend a bit longer on the metaphysics, because at this point in the argument, we have already come close to satisfying some of the “blind criteria for spectrum-hood.” Nature and technology are a genuine metaphysical trade-off. And, as we will now see, this trade-off is entirely distinct from the topics of existing political spectra. We can see both of these points best by looking at the extreme ends of these spectra—the utopias.

4.1 Uniquely Incompatible Utopias

Utopia is a state of affairs where the values emphasized by a given ideology are maximized. Most ideologies (either implicitly or explicitly) posit a utopia. Conservatism of the Burkean/Oakeshottian variety is the exception to the rule as it is defined in opposition to utopian or ideological thinking. Two utopias characterized by locating at either end of the same spectrum, such as an anarcho-capitalist utopia (á la Robert Nozick) versus communist utopia (á la Karl Marx), are incompatible, as one envisages high personal economic liberty, and the other, high shared ownership. But utopias located similarly on one spectrum can differ on orthogonal spectra. It is possible to have a left-wing authoritarian utopia or a left-wing libertarian utopia.

Let's now look at firstly whether technologist and naturalist utopias meet these conditions in each being incompatible with each other, and secondly, whether they are compatible with all four other utopias occupying left, right, authoritarian, and libertarian quadrants. I.e., are all eight octants possible utopias?

We've already shown above the metaphysical trade-off of nature and technology. We might look to Arne Naess or Holmes Rolston III for expressions of naturalist utopia, that is, a world in which maximum intrinsic value is placed on nature. Given the high value placed on areas not touched by human artifice, we would expect much of the world to be in wilderness. For those humans who do exist (for it may be the case that the logical extreme natural utopia contains no humans at all), we might expect some minimal technology to keep them alive, but that is all.

Meanwhile, we could imagine a rough technological utopia as full of sci-fi-esque flying cars, virtual reality, AI, and teleportation. But technological utopias are of course quite hard to imagine *in principle*. Part of technological progress is expanding the art of the possible, many new technologies would have at one point been literally unimaginable. If we knew exactly what to expect, we would have less of a hard time inventing our way there. The key point though is whatever the technological utopia looks like, it seems impossible that it would resemble the naturalists'.

4.2 Compatibility

Theorists across the political landscape have demonstrated that both technologist and naturalist utopias are compatible with other utopias occupying the four quadrants of the left/right and libertarian/authoritarian map.

We can intelligibly posit an eco-communist utopia that rejects technology (e.g. [6]). We can posit one that pursues it, automates work, and equally distributes its

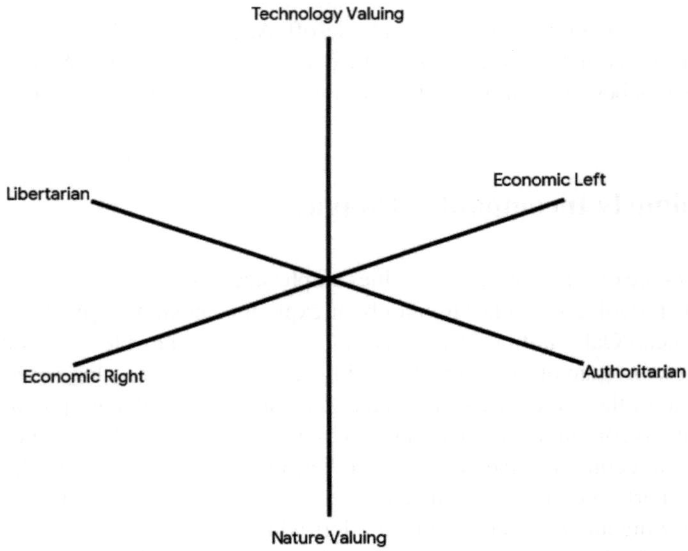

Fig. 4.1 .

benefits (e.g. [1, 7]. We can posit a right-libertarian utopia that respects nature (e.g. [9]), or cyberlibertarians who love technology (e.g. [2, 4]). We can imagine a totalitarian, economically centrist state imposing a pro/anti-nature or pro/anti-tech viewpoint on a populus. And we can imagine an anarcho-syndicalist group of cyborgs, or a nature-loving anarchic leftist commune. To labor the point, if we posited other spectra, such as a spectrum of multiculturalism versus assimilation,these also don't appear incompatible either (Fig. 4.1).

4.3 The Messy Middle

The reality is that most policy decisionshappen in the messy middle, where societies attempt to strike a continuous optimization between values. A form of *prima facie* agreement between sides who intrinsically value nature and technology, respectively, might be around innovations that involve a parsimonious use of technology to facilitate maximal nature. We could imagine a high-tech society that uses technology to minimize their environmental footprint, living in densely packed well-designed cities while leaving large landscapes untouched. Contrast that with a low-tech rural society that burns fossil fuels, harvests too much forest, and so makes a larger, if more rudimentary, physical footprint of humanity on the earth.

It is true that the former society may be doing better than the latter in terms of optimizing some quantity of nature and technology. But this is not to say that such a society will not face continual dilemmas on how and when to opt for more of one or the other. A high-tech society might well dress itself in greenery, but to what extent does it cultivate such greenery or let it grow freely? If the high-tech society truly values naturalness of one sort, that of the surrounding physical environment, then why ought it not, by the same token, value naturalness in human bodies and behavior? And exactly how small should society's footprint be? Is a large city of unnaturalness amid an otherwise wild environment permissible? If only low quantities of urbanity are permissible, then this of course reduces the room within which technologies can be created.

This idea of striking an appropriate balance between competing values implies that this manifests not as a binary choice, but as a continuous optimization. Matters of technological and natural value are not restrictively applied to pure nature and pure artifact but are a background consideration in all matters concerning the manipulation of our physical world. This is the first hint that what we have on our hands is a matter of general disposition or attitude. This, taken with the fact that the attitudes involved are about matters of public policy and public investment, points to the best conceptual apparatus to represent this trade-off as being a political spectrum, and deserved of its place in this three-dimensional plane.

References

1. A. Bastani, in *Fully Automated Luxury Communism: A Manifesto* (Verso, 2019)
2. L. Dahlberg, Cyberlibertarianism. in *Oxford Research Encyclopedia of Communication* (2017). https://oxfordre.com/communication/view/10.1093/acrefore/9780190228613.001.0001/acrefore-9780190228613-e-70
3. S. De-Beauvoir, in *The Second Sex*, 7th edn. (Vintage Books, 1989)
4. N. Land, in *Fanged Noumena: Collected Writings 1987–2007*, ed. by R. Mackay, R. Brassier (Sequence Press, 2012)
5. J. Ortega y Gasset, in *Man the Technician. History as a System and Other Essays Toward a Philosophy of History* (Norton, (1961)
6. D. Pepper, in *Ecosocialism: From Deep Ecology to Social Justice* (Routledge, (1994)
7. N. Srnicek, A. Williams, in *Inventing the Future: Postcapitalism and a World Without Work.* Verso (2015)
8. B. Stiegler, *Technics and Time* (Stanford University Press, 1998)
9. G. Walshe, Green libertarianism. Ethical Theory Moral Pract **17**(5), 955–970 (2014). https://doi.org/10.1007/s10677-0149499-7

Part II
Taking Sides: Arguments for the Value of Nature and Technology

Chapter 5
Nature's Intrinsic Value

Abstract This chapter refines the "Nature" end of the political spectrum by conducting a critique of intrinsic value within environmental ethics, arguing for a necessary conceptual refinement to support the book's core thesis. It begins by contrasting eco-centrism (which intrinsically values nature) with the ambiguous concept of anthropocentrism, which is rejected as an analytically empty term that fails to provide a substantive opposing political value. The chapter proposes a vital distinction in the intrinsic value of nature, cleaving the concept along the lineages of its two forefathers: Moorean Intrinsic Value and Kantian Intrinsic Worth. Moorean Value concerns the abstract, non-person-related value of "naturalness" as a state of affairs (e.g., Robert Elliot's Faking Nature), which is identified as the relevant intrinsic value for the Nature-Technology spectrum and the proper subject of this book. Conversely, Kantian Worth concerns the value of biota as moral persons (e.g., Peter Singer's sentience-based utilitarianism and Tom Regan's rights-based deontology). The chapter demonstrates how these two values are not only distinct but can be actively opposed (e.g., animal welfare goals versus naturalness goals), clarifying confusion within environmentalism.

Keywords Environmental ethics · Intrinsic value taxonomy · Moorean intrinsic value · Kantian intrinsic worth · Anthropocentrism critique · Eco-centrism · Animal rights · Naturalness versus personhood · Sentience · Environmental policy trade-offs · Synthetic biology

The main cleavage in the field of environmental ethics is that between eco-centrists (or bio-centrists, or "deep ecologists"): those who intrinsically value nature, and anthropocentrists (or "shallow ecologists"): those who have a more pragmatic appreciation of the instrumental value of nature, but don't go as far as to say its value is intrinsic.

Philosophers, economists, and policymakers alike recognize that parts of nature are instrumentally valuable to economic welfare (narrowly conceived). Nature offers us raw materials, food, human habitats, tourist income generators, and providers of other ecosystem services like carbon sequestration. Less obviously, and depending on one's ethical framework, Nature might be instrumentally valuable in other ways

B. Steyn, *Technology vs Nature*,
SpringerBriefs in Applied Sciences and Technology,
https://doi.org/10.1007/978-3-032-18582-2_5

to other intrinsically valuable ends. Nature inspires scientific endeavor, and so is an instrument and substrate for knowledge; it offers mountains and rocky crags that are instrumental to leading a life involving an intrinsically valuable "adventure," and might provide instances of tranquil and quiet sanctuary, an instrument to the intrinsically valuable "peace and security." These features alone are enough to buttress strong arguments in favor of the conservation and proper treatment of the natural environment.

But over and above the instrumental benefits that might occur, the eco-centrists think nature (or parts thereof) might be bearers of value for its own sake. This gives an additional *pro tanto* argument in favor of policies like conservation or refraining from exploring certain types of innovation.

Environmental ethicists' definitions of anthropocentrism and eco-centrism hinge on whether or not one embraces the intrinsic value of nature as part of "the good." Some theorists would say anthropocentrism is an ethical stance that "the only good that matters is the good for humans," ruling out a good for non-human nature. But others, whom we might call "enlightened anthropocentrists," contend that while anthropocentrism is a true *metaethical* position, because morality is a human construction, concerned with the good of humans only, it is still nevertheless possible that preserving nature constitutes part of the intrinsic good for humans.

In this metaethical sense then, anthropocentrism is an empty term, as all it says is that value considerations ought to be the considerations of humans as the valuers, but not that the *content* of the values need to be about humans. To preserve its intended derogatory slant, anthropocentrism ought to mean both that the only good that matters is the good for humans, and, furthermore, that nature isn't among the intrinsic values that constitute the good for humans. Similarly, eco-centrism could be construed as either the metaethical stance that there is a good for non-human nature, or that nature constitutes part of the human good.

Whichever way it is defined, ethically or metaethically, this distinction is broadly helpful in explaining divergent attitudes in environmental policy, and, in particular, in separating intrinsic value considerations from instrumental value considerations. Using this distinction, we can give an account of why the environmental activist campaigns against GM crops, nuclear power, geoengineering, etc., even when they accept the instrumental (e.g., economic) arguments in favor of these activities—they want to protect *nature*.

However, the explanatory power of the anthropocentrist/eco-centrist distinction is limited. Eco-centrism and anthropocentrism do not constitute either side of a political values spectrum. Anthropocentrism is only defined as an *absence* of a particular intrinsic value for some non-human entity. An absence that could imply either ambivalence or hatred of that entity. It cannot distinguish between someone who hates nature and loves its opposite, and someone who is just not that bothered. There is no positive intrinsic value on the anthropocentrist side. Anthropocentrism is not conceptually unique. We could similarly take any intrinsic value X and create a concept to mean "not an X-lover." Being "not an x-lover" is not a substantive point of view. By analogy, consider the economic spectrum: If one is not bothered about equality, they could be labeled a "non-socialist," but this is not a philosophically charitable or useful description. The person's view isn't defined only in opposition to socialism, but rather in support of something substantive, which, as it happens, directly contradicts socialism.

Unlike the vacant idea of anthropocentrism, the political value spectrum we are building in this book can more uniquely supply the opposing intrinsic political value to eco-centrism. While the eco-centrist might accuse the environmental economist of ignoring the intrinsic value of nature, and only concentrating on its instrumental value; those who intrinsically value nature do not routinely consider the negative consequences of their position. They view those who don't value nature as just cruel, soulless, ruthlessly calculating creatures. But, in fact, they may have an equally vibrant soul but of a different sort. It is a soul which admires and perhaps romanticizes human creativity and humanities' bettering of nature—we will turn to that side of the debate later.

But first, what arguments do the eco-centrists make? Robert Elliot's "Faking Nature" example earlier gives us a nice way to get into the headspace of intrinsic value for nature…it's the origin of the woodland that matters, the fact it is natural. Not its aesthetic or any other instrumental benefit. Richard Sylvan (née Routley) gives us another thought experiment. Consider the "last man" on earth in a post-apocalyptic future. He happens to be a man who commits gratuitous harm to flora and fauna. Routley intends the reader to intuit that it is wrong for the Last Man to harm nature, even if there are no instrumental human consequences, ergo, nature has intrinsic value [7]. Others have offered similar arguments: Holmes Rolston III considers intrinsic value for the African Butterfly not to be killed by a collector [6], and Robin Attfield considers a specific "last man" scenario on whether he should chop down a tree, positing intrinsic value for the tree.

Unfortunately, the concept of intrinsic value in environmental ethics is rather vexed. Commonly in philosophy, ongoing debates about the fundamental nature of concepts don't stop those in more applied settings wielding the concepts in question. Environmental ethics boomed in the 80s and 90s, with many making claims about the intrinsic value of nature or parts thereof. And much of this work precedes major developments in axiology. As a result, much of the literature plays relatively fast and loose with the concept of intrinsic value.

If there is this intrinsic value for nature, how does one weigh it against instrumental value when we make policy decisions? Controversially, but perhaps pragmatically, some environmental economists seek to proxy this extra value by teasing out figures for the public's "willingness to pay" for an environmental good, as if it were a market good. For instance, non-market valuation methods have been used to generate a willingness-to-pay measures for threatened, engaged, and rare species [8].

Many think this type of endeavor makes a category mistake. A "willingness-to-pay" methodology is a confused attempt to draw out what is actually a property that is *irreducible* to economic value. It is equivalent to me asking: "how much would you be willing to pay to buy your own dad?" While it is true that in, say, a hostage or ransom situation, one may be forced to throw out a figure, one might think that this is nevertheless not sufficient. In fact, the answer is just indeterminate because we wouldn't think it right for me to put a price on him. Approaching environmental policy this way nevertheless gives some non-zero valuation, which is perhaps needed in a world of public choice decision-making which is, for wider institutional reasons, reliant on reductionistic cost–benefit analysis models.

But there are also more fundamental concerns about the concept of intrinsic value itself. In a 2024 paper in The Journal of Environmental Ethics, I taxonomized the ways in which thinkers have thought about nature's intrinsic value [12]. I adopt more recent innovations in axiology, notably those of Bradley [2], building on Korsgaard [3], to precisify the concept of intrinsic value as it applies to nature, cleaving apart the concept along the lineages of its two forefathers, G.E. Moore and Immanuel Kant.

The crucial take away from that paper is this: When we say that X (be it nature or some other X) is intrinsically valuable, valuable "as an end in itself," this is ambiguous. It could mean one of two things:

(1) Moorean Intrinsic Value: X is a feature of some state of affairs that the valuer believes should be brought about in principle, fundamentally, for its own sake, because it makes the world a better place in virtue of its existence and to the degree that it exists, not merely as an instrument to some further end. For instance, "truth," "beauty," "honor," and "love" are classically held to be abstract properties that are intrinsically valuable, part of the good life for their own sake.
(2) Kantian Intrinsic Worth: X is a moral patient, a subject of moral consideration, something to whom we owe deontological duties, rights, or include within a consequentialist calculus, depending on one's moral framework. The most common inputs for X here are human beings: Kantian ethics starts from the basis of valuing each human life as an end in itself.

Nature, the natural environment, and parts thereof, has been argued by different thinkers to be intrinsically valuable in both of these ways.

For some, naturalness itself is a property of the world that is valuable and should be protected. Take, for instance, Robert Elliot, with his arguments about a restoration of a woodland being tantamount to forgery. Or, as we will see later, take human enhancement thinkers who desire to protect human nature, precisely because it is natural and for no other reason.

Within this realm of discussion are arguments that certain properties of natural environments can be value-intensifiers, making them even more valuable, for instance, those that display ecosystem integrity, biodiversity, complexity, and rarity. There is also much discussion about aesthetics in nature. Perhaps both artificial and natural environments can exhibit beauty, but some want to argue that when nature achieves beauty, there is something additionally valuable there, not only in the superficial beauty itself, but in the fact it was naturally produced.

This Moorean variety is the sort of intrinsic value for nature that is properly the subject of this book. But others in environmental ethics are undertaking the Kantian project; they want to extend moral consideration to non-human biological life.

Take, for instance, Tom Regan, who, in "The Case for Animal Rights" [5], advances a deontological principle that those who are a "subject-of-a-life" (criteria for this include psychological continuity, sense-perceptions, motives) have a moral right to respectful treatment and not to be harmed. Expanding the circle of concern from fauna to flora, in "The Good of Trees" [1], Robin Attfield advances that all beings which can be harmed or benefited, all which can be said to "flourish" or have a good for it, can hold intrinsic value. Or, most famously, Singer [11], a utilitarian, famously advances that all sentient creatures are morally considerable beings, as they are recipients of pleasure (utility) and pain (disutility). Excluding them from the moral calculus is, in Singer's view, arbitrary. Singer's position is like that of Regan's, Atfield's, and others discussed above in that he considers animals as being deserving of some sort of ethical consideration as persons. Singer sees animals as "ends in themselves" in the sense that they are bearers of utility.

As my paper argues, in the twenty-first century, we need to be more precise about our desiderata. We need to more sharply distinguish between the Mooreans and the Kantians. I.e., on the one hand, the nature lovers, and on the other, the animal rights lovers.

At this point in history, observing the distinction between life and "natural life" hasn't mattered, as most of life crosses the threshold point for us to loosely consider it natural, but with the rise of synthetic biology, extensions of genetic engineering (beyond the age-old artificial breeding), and perhaps one day, sentient artificial intelligence, this may not soon be so. Sandler [9] considers whether attributions of intrinsic value can be extended to synthetic biology. He argues that, essentially, while synthetic biology might have less intrinsic value with respect to it having lower naturalness (i.e., Moorean), the "value for biota as persons" (i.e., Kantian) views are unaffected.

There is clearly a "good for" beings moving along the natural/artifactual scale. Sandler gives an example of breaking the legs of a dog (a semi-artificial being) being bad, and the fact that there is such a thing as a "good for" a purely synthetic protocell. It seems arbitrary then for the thinkers of this section not to extend their moral considerations to these sorts of beings. For sentientists like Singer and Regan, the justification for valuing animals stems from cognition, which artificial life can have. For Conativists like Attfield, it hinges on life, which artificial life, by definition, can have.

We should see a clear distinction between the Moorean endeavor of valuing abstract states of affairs concerning nature, and the Kantian ethic of valuing biota as persons. It is logically possible to strongly hold convictions about the former without seeing parts of nature as persons at all, but, moreover, treating biota as persons may actively work against the goals of the nature-lover. It might be the case that what is good for a biological entity involves minimizing human interference with it (i.e., increasing its naturalness), but it need not be. Animal welfare considerations could well lead us to reduce naturalness in certain circumstances. After 1000s of years of being bred as human companions, dogs are domesticated animals whose welfare is likely to be improved in human environments rather than wild ones. Moreover, we currently intentionally breed farm animals for their docility and slovenliness—traits which would not serve them well if released back into the "natural habitat" of their closest wild evolutionary ancestors.

One response could be that humans have already doomed these species, and we should only focus concern on wild animals. But animal welfarists of the types discussed above are generally concerned with the welfare of domesticated animals and also have no way to demarcate domesticated animals from non-domesticated as units of moral concern. They are still sentient, and it would be arbitrary to disregard them and not to try and better their circumstances. Moreover, recent estimates have suggested that most of the mammalian biomass on earth is livestock. Ignoring livestock therefore seems very ignorant indeed. An animal ethic focused only on wildlife is almost irrelevant in the twenty-first century.

Nor is it clear that natural habitats are preferable even to wildlife. In *The Hedonistic Imperative*, Pearce [4] argues that our moral duty to avoid suffering in humans and animals means we ought not only to prevent artificial forms of suffering but actively innovate with technology to alleviate a natural baseline of suffering in the wild. He suggests we should aim to "reprogram predators" to protect prey, and "civilize the animal kingdom." As Pearce points out, there is nothing in an ethical framework that values animal preferences which rules out such an approach, and consistency demands it be taken. Needless to say, this would be anathema to naturalists like Elliot.

Moreover, treating individual biota as persons can be problematic for important system-level biological properties like biodiversity and ecosystem integrity in circumstances where an invasive species is threatening an incumbent population. Sarkar gives a good example of this tension; a case of stray dogs threatening to eat the eggs of a rare turtle [10]. He criticizes Singer's account as entailing the wrong outcome to this scenario, which would be to refrain from culling the dogs and let them run rampant, damaging the naturally occurring, albeit less sentient, unborn turtles.

In distinguishing between the value of nature and non-human personhood, we can see once again that the term "Anthropocentrism" is misleading and ill-defined. One might think that expanding moral concern to animals is plausible, but see this as of-a-piece with expanding moral concern to artificial life. If "anthropocentric" means "doesn't value nature intrinsically," then it is a misnomer as an anthropocentrist can nevertheless hold non-anthropoids to be morally central. Put another way: not respecting AI as a moral patient could equally make one an "anthropocentrist."

In distinguishing between concerns that lie on the nature/technology spectrum, and concerns about the personhood of non-human beings, we can explain the tensions within ideological groups that *prima facie* share a common goal. For a fringe movement, indeed for any political movement, it might be practical to make ideological concessions to expand the group's size. But in generations to come, debates around nature and technology will become a greater feature of our politics and so the small distinctions matter more.

There are many debates within environmental ethics which do not make sense if we treat these diverse interests as a coherent group of "environmentalists." Environmental groups already exhibit heated internal debates about subjects like culling, and whether we ought to focus efforts more on conservation or artificial regeneration of habitats.

But upon the advent of artificial sentient life, Greenpeace might radically cleave apart when they realize that half their supporters are beholden more to naturalness, while others want to expand the moral circle to all sentient life. Animal rights activists in the mold of David Pearce might find themselves championing the bleeding edge of technological development to tamper with their brains to increase base levels of serotonin and dopamine, or in civilizing predatory nature. Others will want to simply leave animals be. The distinctions made here allow us to navigate our way through the confusion.

With nature's intrinsic value now more clearly defined, and distinguished from adjacent ethical theories, we now turn to the other side of our spectrum.

References

1. R. Attfield, The good of trees. J. Value Inquiry **15**, 35–54 (1981)
2. B. Bradley, Two concepts of intrinsic value. Ethic. Theory Moral Prac. **9**, 111–130 (2006). https://doi.org/10.1007/s10677-006-9009-7
3. C. Korsgaard, Two distinctions in goodness. Philosophical Rev. **92**, 169–195 (1983)
4. D. Pearce, in *The Hedonistic Imperative* (2004). http://www.hedweb.com/hedab.htm
5. T. Regan, *The Case for Animal Rights* (Routledge, 1988)
6. H. Rolston, Is there an ecological ethic? Ethics **85**, 93–109 (1975)
7. R. Routley, Is there a need for a new, an environmental ethic? in *Proceedings of the 15th World Congress of Philosophy*, vol. 1 (Sophia Press, 1973), pp. 205–210
8. M. Sagoff, in *The Economy of the Earth: Philosophy, Law, and the Environment*, 2nd edn. (Cambridge University Press, 2007). https://doi.org/10.1017/CBO9780511817472
9. R. Sandler, Is artefactualness a value-relevant property of living things? Synthese **185**(1), 89–102 (2012)
10. S. Sarkar, in *Biodiversity and Environmental Philosophy: An Introduction* (Cambridge University Press, 2005)
11. P. Singer, in *Animal Liberation: A New Ethics for our Treatment of Animals*. New York Review (1975)
12. B. Steyn, Nature's intrinsic value. Environ. Ethics **46**(2), 107–130 (2024)

Chapter 6
Technology's Intrinsic Value

Abstract This chapter provides the crucial, positive counterpoint to the intrinsic value of nature, establishing the Intrinsic Value of Technology as the opposing pole of the political spectrum. It begins by arguing that technology's value extends far beyond its obvious instrumental and economic benefits, positing that a deeper "technological imperative" exists, valuing technological states of affairs *suis generis* (for their own sake) under a Moorean conception of intrinsic value. Since this position is largely undefended in analytical philosophy, four distinct lines of argument are advanced: (1) Brute Intuition, using examples like spacecraft to evoke direct, non-instrumental appreciation; (2) Appeal to Liberty, arguing that functional artifacts uniquely expand physical liberty (or Amartya Sen's "option sets"), which is a political value on par with social and economic liberty; (3) Appeal to Virtue Epistemology, drawing an analogy between the value of knowledge and the value of technology, where the creation of artifacts is defended as a "technological achievement" that is credit-worthy, often as an irreducibly collective human virtue; and (4) Appeal to Misattributed Value for Scientific Knowledge, demonstrating that much of the public and philosophical esteem for "science" is in fact owed to the concrete act of engineering (the application of knowledge) as exemplified by record-breaking technologies and the Technology Readiness Level (TRL) spectrum.

Keywords Intrinsic value of technology · Philosophy of technology · Technological imperative · Moorean intrinsic value · Physical liberty · Amartya sen · Option sets · Virtue epistemology · Technological achievement · Engineering versus science · Technology readiness level (TRL)

Sure, nature's alright…but what about *technology*?

We can take piles of silicone, metal, and rock, and weave them together in such a way that I can see my friend's face rendered on a screen and talk to her despite a Coronavirus Lockdown. I walk through history museums admiring the huge variety of tools made by ancient peoples to achieve survival, beauty, and pleasure. More mundanely, I admire the smoothness with which soft-close cabinets shut, the efficient parsimony of a ring-pull on a Coke can, the resourcefulness of the modern chicken

B. Steyn, *Technology vs Nature*,
SpringerBriefs in Applied Sciences and Technology,
https://doi.org/10.1007/978-3-032-18582-2_6

nugget as a use for the otherwise inedible, and the way door handles are designed intuitively so one knows whether they are to be pushed, pulled, twisted, or turned.

Technological pursuits very obviously have instrumental value, i.e., value as means to some other valuable ends (e.g., utility, health, comfort, the pursuit of scientific knowledge). But I will argue that furthermore, they have *intrinsic* value, i.e., value as ends in themselves. Recall, here I adopt a broadly Moorean conception where intrinsic value means that some state of affairs is valuable for its own sake and is subject to the maximization principle (i.e., more of it makes the world a better place all else being equal).

The position that technology has intrinsic value has not been defended before in analytical philosophy (in later chapters, we will turn to some thinkers who have touched on the subject in the continental tradition). I think there are three main reasons for this: First, analytical philosophy of technology is a young and disparate field. Second, technology's *instrumental* value is so obvious that proponents of technological progress probably only need to arm themselves with those to advance their cause. And third, technology tends to command a high economic valuation and so is not underappreciated by free markets relative to, say, nature, the arts, or some other political or ethical good. Unlike nature, and somewhat unlike art, technology is often created with the explicit intention to make money.

However, there doesn't seem to be a relevant difference in principle. We pay for art and knowledge, and we pay to experience or conserve nature, though we don't take these concepts' value to be just that which we pay. There is a deeper technological imperative that exists in at least some of us, a desire to push the frontier further, that is not reducible to narrow welfare considerations. In what follows, I consider various lines of argument leading to the conclusion that technological states of affairs are *suis generis* intrinsically valuable. My aim here isn't to pick and choose between them, and they need not be mutually exclusive. It is rather my hope that for each line I can provide enough of a foundation to start a conversation, which so far doesn't seem to be happening at all.

6.1 Argument 1: Brute Intuition

Spacecraft are incredibly cool. As far as I can tell through introspection, my appreciation of the spacecraft is not shaped by a narrowly conceived instrumental reasoning that they will *serve me.* My appreciation of the spacecraft doesn't stem from an ambition of mine to go to space, nor particularly from what I perceive to be spillover benefits to other industries and the economy. These instrumental benefits might well obtain but are not responsible for my actual state of mind, which is much more powerful and directly appreciative of the endeavor, in a way I struggle to know how to characterize precisely other than saying: I appreciate it for its own sake.

I want to live in a world where people are endeavoring to build things that allow us to explore the wider universe. I intrinsically appreciate this state of affairs, and I admire engineers who have dedicated their lives to helping to realize it. I could say the same about many other technologies and technologists innovating in areas like architecture, artificial intelligence, machine learning, synthetic biology, agriculture, fashion, and gastronomy.

For me then, there is a lot that is intuitively admirable about technological endeavors. But like in other cases of candidate intrinsic values, it is difficult to directly argue for the conclusion that this admiration logically entails value. It may be enough to just baldly assert that technology or engineering has intrinsic value, perhaps using illustrative case studies like spacecraft to induce the right intuitions. You could simply intuitively agree or disagree, and that is that.

Some may want to explain away these examples as being valuable for one reason or another. Maybe it is a covert form of instrumental value truly at play? A common strategy to try and show the presence of intrinsic value is to isolate the very thing that is in question and ask if it is valuable. In our case, this would be looking at examples of technology that could be seen as purely technology for technology's sake. They exist only as instantiations of technological excellence, with no instrumental value. A literal and figurative "toy example" is a Rube Goldberg machine, named after the eponymous American cartoonist, which is a chain reaction contraption made to perform a simple task in an overly complicated and indirect way, famously used by Wallace in the stop-motion cartoon Wallace and Gromit. Every year, there is a Rube Goldberg machine competition, where there is a large community of professional builders and junior entrants who revel in the artistry, comedy, and insanity of ever-more convoluted machines.

Watching a Rube Goldberg machine is interesting, fun, and marvelous. It might push similar buttons in you as watching an artistic performance. It might be right to call the appreciation aesthetic, but even if it is aesthetic, it is not simply that what I am watching is a sensory pleasure. The appreciation seems to stem from thinking about the craft and engineering behaviors involved, and the fact that it doesn't merely set off a bunch of unconnected mechanical processes, but rather, they all connect intimately such that one can say about the machine that it works. Carmaker Honda, whose 2003 advert Cog won many awards, appropriated the Rube Goldberg concept. The voiceover said, "isn't it nice…when things just… work." "Yes," would be my reply. And although I may part with cash to "own" this feeling of things working, the fact that I do does not preclude from this feeling being something intrinsically valuable.

This sort of appeal to pro-technology intuitions, which is par for the course in the enemy camp of environmental ethics, is limited by my own abilities as a creative writer. Maybe they have some force nevertheless, but I think there are also better philosophical arguments that can be made.

6.2 Argument 2: Appeal to Liberty

Earlier, we employed a definition of technology as "intentional functional artifact." *Prima facie,* it may seem odd to assert that something that is, by definition, *functional* can be intrinsically valuable. Functions are instrumental by definition. They are means to an end. Functions are defined with respect to particular ends. The function of scissors is to achieve the end of cutting thin objects (paper, cardboard, plastic). Is it not the case then that all functions are only instrumentally valuable to the ends they serve, and so the value of the artifacts that bear this functionality is only instrumental?

I think the reduction of the value of functional artifacts to only instrumental value may miss something important. What if the very presence of functionality is itself intrinsically valuable. To see this, we first need to make a detour into the political philosophy of liberty. Consider this passage from Isaiah Berlin.

> If, as I believe, the ends of men are many, and not all of them are in principle compatible with each other, then the possibility of conflict—and of tragedy—can never wholly be eliminated from human life, either personal or social. The necessity of choosing between absolute claims is then an inescapable characteristic of the human condition. This gives its value to freedom as Action had conceived of it—as an end in itself, and not as a temporary need, arising out of our confused notions and disordered lives, a predicament which a panacea could one day put right. Berlin [2, p. 54].

Berlin here is outlining the powerful idea that liberty has intrinsic value, not as a "temporary need," i.e., an instrument toward some settled state of living, and even despite the fact that this liberty may create conflicts and tragedies.

Amartya Sen progresses this line of thinking across different works using the concept of an "option set"—a set of options that are feasible for a person to do in some situation that are jointly exhaustive and mutually exclusive in that one must end up selecting one or other of the options. Sen gives a thought experiment comparing two scenarios, in one, somebody is starving but has no choice in the matter. In the other, the person chooses to starve for religious reasons—but could have chosen otherwise. Both end up in effect with the same life option, but as the latter's option set is wider, their situation is better [20, p 201].

Over and above the *content* of the choices people make, there is intrinsic value in the very fact that they have a choice. The size of one's option set is itself a source of intrinsic value.

Returning now to our subject matter, what should be clear is that functional artifacts expand our liberty (or, as Sen puts it, option sets,) in a unique way. Functional artifacts give us physical freedoms to manipulate and negotiate the world around us: The freedom to fly, to examine objects in microscopic detail, to arrange large objects, to explore virtual worlds, to play with simulated characters, to order a takeaway from one's phone, to crowdfund a new business venture, and to learn a new skill. With only the functions available to us through our bare bodies, our physical option set is limited. But with functional artifacts, our option sets vastly expand.

It may not be so easy to see a meaningful expansion of option sets at the level of an individual instance of an artifact. But the expansion is much clearer when considering aggregated technological states of affairs. Consider two worlds: (1) a hunter-gather style world with only minimal and basic functional artifacts and (2) the United Kingdom in the present day. It is clear that the latter world has a vastly expanded physical option set, and this comprises many instances of many different technologies. And this, over and above the particular instrumental value of the technologies present, is a valuable state of affairs, and one to build on.

Sen, articulating his view about which expansions of option sets actually increase value, makes a throwaway analogy to having two cars. He notes that it surely wouldn't increase the value of the situation much or at all if one had two identical cars in their garage, one with a slightly defective gearbox. Yes, it is an additional option, and it might still be a good one, but not one that substantially improves one's opportunities. He goes on to precisify which option sets truly offer additional opportunities [21, p. 531].

The details of Sen's argument don't matter here. What is telling for us is Sen's use of the analogy of a car. In this particular case, the second car gives no meaningful expansion to his option set. But the argument presupposes that certain states of affairs in the vicinity to the one posed can be evaluated positively, e.g., if the second car genuinely diverged enough from the first to increase the option set, or indeed having a car at all versus having to walk. Sen is only using this as an analogy to his true desiderata, political and social liberty. But if we reverse the analogy, we can see Sen implicitly appreciated physical liberty here as a coherent source of value.

Looking beyond Sen, some aspects of traditional political liberal thought do already touch on physical liberty. We are concerned with questions of bodily autonomy, freedom of movement, and the rights of citizens not to be incarcerated. But discussion has historically been limited to one's use (or lack thereof) of their own body, and not particularly of the use and functionality of artifacts. That said, there is now a growing discourse in international development policy around whether access to the Internet and digital technology is becoming a "human right." Implying that this technological liberty, like the liberties of private property, free speech, and free association, is a core part of the good life.

Of course, one might dispute that liberty of *any* kind is intrinsically valuable, saying that any value one attributes to liberty is ultimately instrumental. For instance, saying that the freedom to access the Internet isn't inherently good, but just good because it brings about particular good effects such as social connection, commerce, and entertainment. One might even go further, in parallel to Communitarian critiques of liberalism, and say that physical liberty is intrinsically *harmful*, as we ought to be trying to live similar lives with one another, harmonizing our physical option set around a set of activities rooted in tradition and cultural capital. Of course, the counterarguments are compelling in many areas. But I am happy to be partners in guilt with the other planes of liberty here. I am merely trying to show that physical liberty is on a par with social and economic liberty, in that for at least some, it could validly be taken to be an intrinsic value unto itself.

In reality, all liberalisms face internal paradoxes. The liberty of certain individuals may curtail the liberty of others. Unrestricted physical liberty for some, as with unrestricted economic and social liberties, could create hegemonic technological societies which restrict the technologies available to others. And yet, this physical liberty is something one might still reasonably think is part of a good life for all when held to some degree.

There is a technical point here. Have we shown an argument for technology being intrinsically valuable? Not quite. All we have argued for is that physical liberty is intrinsically valuable. Functional artifacts are just an *instrument* to the intrinsically valuable physical liberty. A means to achieve the end of a wider option set. But in practice, these positions are barely distinguishable.

Some modicum of physical liberty is realizable without technology. We can achieve things in the physical world using our bare bodies and enlisting the help of others and their bodies. For instance, I am free to experience what it is like to be 20 feet in the sky by being at the top of a human pyramid. It might be added that nature too is a source of physical liberty. Another way to get 20 feet in the sky is through climbing a mountain, so long as I don't use a piece of bamboo as a walking stick (thereby creating a technological artifact).

But beyond the confines of our bare bodies and the few affordances given by the untouched natural environment, technologies are effectively *the only* means of expanding our physical liberty. We can achieve physical liberty beyond our bodies to the extent that technologies are present, and only in the ways circumscribed by the particular technologies that are present. Nail clippers serve the function of clipping nails. This additional functionality is the way our physical liberty expands. States of affairs that contain large amounts of physical liberty will very likely contain technologies, and vice versa. Physical liberty supervenes on technology.

So, in summary, while we have not precisely shown functional artifacts themselves to be intrinsically valuable, we have identified intrinsic value reliably in the vicinity. Technologies are the sole source of physical liberty beyond the bare human body, and this liberty, an expanded option set, is intrinsically valuable. But in political terms, this cashes out the same: an intrinsic motivation toward technological advancement.

In our efforts to show that technological progress is a *political* value, it should encourage us to observe that it shares something in common with other political values, namely, that each is concerned with a different plane of liberty (social, economic…and here, *physical*). Maybe it is possible to argue that a distinguishing feature of political values is that they *all* concern some type of liberty or lack thereof. But this meta-axiological point is not my concern in this book, and, as we are about to see, there might be other ways to buttress the intrinsic value claim without appealing to liberty.

6.3 Argument 3: Appeal to Virtue Epistemology

Contemporary virtue epistemology is a broad collection of theories focused on the normative importance of intellectual virtue. Ernest [23] was first to introduce into modern philosophy the notion of an intellectual virtue, arguing that it can resolve a longstanding conflict between coherentist and foundationalist epistemology. Intellectual virtue concepts have been employed to tackle Gettier problems, the internalism/externalism debate to skepticism, and more. Since Sosa, there has been a split between reliablists, who tend to see intellectual virtues as certain cognitive faculties or powers like memory and introspection, and responsibilists, who see the virtues as character traits involved in knowledge production, such as inquisitiveness, fair-mindedness, and intellectual carefulness.

Virtue epistemologists advance that the exercise of intellectual virtues, like ethical virtues, has a normative value. A key area of enquiry for virtue epistemologists is the Meno Problem—why is knowledge more valuable than mere true belief? Zagzebski [6], Greco [7, 28] and Riggs [16–18] and Sosa [24, 25] all attempt to solve the Meno problem using the notion of "credit." Knowledge is more valuable than true belief because knowledge is a credit-worthy state of an agent. The successful exercising of one's intellectual virtues is an achievement and that achievement is valuable over and above the true belief itself.

Recall that when we defined technology, we gave a tripartite definition "technology = intentional functional artifact," which is structurally similar to "knowledge = justified true belief." Now suppose that there is an equivalent "Technology Meno Problem": why are "intentional functional artifacts" more valuable than mere "functional artifacts"?

Equivalently to virtue epistemologists, we can respond to the Technology Meno Problem by arguing that the creation of functional artifacts through intentional action is *credit-worthy*. The successful intentional creation of functional artifacts is an achievement. It is not an epistemic achievement as it does not relate to knowledge production, but rather a distinct type, "technological achievement," relating to functional artifact production.

Following the analogy through, we can see a set of character traits that dispose someone to create/improve/promulgate technology as potentially being candidates for virtues. Paralleling the split among virtue epistemologists, we can consider "reliabilist" technological virtues, for instance, the exercise of mental faculties involved in wielding and combining propositions and know-how to form technological concepts, or, at a later stage of the production process, the skill of a factory worker in executing the production of a component to an exact specification. Or, as "responsibilists," we could consider certain character traits in technologists to be virtuous, such as carefulness, exactness, and attention to detail.

There is no reason not to see technological achievements as credit-worthy *on a par* with intellectual achievements. Some of our technological achievements are wondrous. From early hominid's innovations in toolmaking, to the printing press, to the Internet. But we do not have to refer to these grandiose examples. A great many more technological achievements are less wondrous, but still locally credit-worthy in the life of an average individual. A credit-worthy exercise of one's technological virtue could be as simple as making a dinner, decorating one's room, making art, doing DIY, or knitting a scarf.

In the modern world, I can conveniently and cheaply eat meals prepared by someone else. I can hire tradespeople to affix my new shelving unit. I can simply buy art and scarves of better quality than I can make myself. And yet, a reduction of the value of these activities to just the functional artifacts themselves misses the point. I find it no surprise that it remains a source of personal pride and social respect for people who cook their own food and build their own shelf even when they can conveniently achieve these ends in other ways. It is because exercising one's technological virtues is part of a life well lived.

DIY and home-cooking are exceptions to the norm in the modern world. Our modern way of life does not place central importance on technological virtue for all people. Rather, we outsource technological production to a concentrated minority of experts: engineers and manufacturers of various kinds. This makes it an area of virtue that is less accessible, and perhaps why it is less the subject of axiological attention. Commonly cited moral and epistemic virtues feature heavily in all of our lives. We are all confronted by regular opportunities to practise honesty, love, friendship, curiosity, etc. But it is more possible to lead a life completely lacking in any technological virtue.

We might consider that appreciating virtue in others is a virtue unto itself. I admire the great technological achievements of others, past and present. From early hominid's innovations in toolmaking, to the printing press, to the Internet, to the latest generation smartphone.

But we should also see that, with a broader view of things, we do collectively play a part in technological production. With all of these examples of highly credit-worthy technological production, it is very difficult to attribute the achievement to any single individual.

So how then do we apply the notion of "credit," given the complexity? How do we deal with knowledge (and hence technology) as a collective human achievement? A key line of criticism of the credit-worthiness approach to virtue epistemology is that many instances of some agent having knowledge don't in fact seem to be credit-worthy at all for the agent in question. If we cannot attribute credit, this mounts a potential challenge to knowledge being valuable. And if, analogously, we cannot attribute credit for technological endeavor, this mounts an equivalent challenge to my argument.

Lackey [12] gives an example where a man called Morris, new to the city of Chicago, asks a passerby for directions. The passerby gives great directions, and on that basis, Morris forms a true belief. Here, while Morris has knowledge, it was the passerby's contribution that was integral to explaining why Morris learned the truth. Morris doesn't deserve any credit.

One line of response from Riggs [18] is to reject that Morris has knowledge in the first place. But trying to rule this out rules out a lot of knowledge derived from testimony that we would intuitively want to call knowledge. The consensus is that credit is present, but that it is in some way spread across a collective. One line here attempts to redeem Morris as still individually credit-worthy for his contribution to the successful instance of knowledge dissemination. Greco [5] argues that he deserves credit for "having learned the truth." Morris has no guarantee that the passerby is dependable. Morris exercises various cognitive abilities, based on a complex social intuition, to decide to accept the passerby's testimony. That deserves credit.

Sosa [25] similarly argues that Morris deserves partial credit for his role in the collective endeavor of knowledge transmission. He uses the analogy of a quarterback in an American Football game who throws a touchdown pass: *"The quarterback's pass derives from his competence, but its great success, its being a touchdown pass, manifests more fully the team's competence."* Similarly, the intellectual achievement of "knowledge of Morris's whereabouts in Chicago" is not attributable to Morris alone, however, like the runner who catches the winning pass from the quarterback, Morris has "played his part" in receiving it aptly.

But a further line of response, taken up by Greco in later work [8], is to attribute epistemic virtue not to individual knowers at all, but to broader intellectual communities. Greco develops an account of "joint agency." In the case of the achievement of testimonial knowledge, it neither lies with the speaker (passerby) nor the hearer (Morris), but rather, the competent joint agency of both acting together.

Technologies are, even more so than knowledge, irreducibly collective achievements. And like knowledge and sporting teamwork, this is part of what makes them particularly valuable. Our technologies instantiate the collaborative efforts of many agents working in harmony. Technologies are a testament to our ability to communicate and overcome the limitations of our lone selves, and they are the fruits of complex social endeavors, carried over many generations through teaching and culture, in a way that separates humanity from other species.

Like the "touchdown pass" and the successful transmission of knowledge through testimony, many technological virtues are exercised socially. And so, their credit is not attributable to single individuals. But they do seem credit-worthy nevertheless. The successful sharing of a set of intentions and goals, being in sync, are phenomena that amount to something of value, over-and-above the value of the end product itself.

For more quotidian examples, consider the unique experience of cooking together with a loved one, or the parent and child working together in a garage to build a soapbox racing car. The value in these cases is neither reducible to the output (the meal/car), nor some side benefit, (e.g., the opportunity to merely spend time with one another, or ease boredom). Rather, there is a special significance in the co-production of an artifact. It is a social achievement, a distinct form of human bonding.

So, in summary, the distributed and complex nature of assigning some instances credit is not a challenge but rather a strength of a credit-based account of technological value. It allows us to accommodate intuitions in both the cases where there is individual credit and socially distributed credit.

Lackey [12] raises another issue for credit-based accounts of the value of knowledge. In those instances of knowledge that can straightforwardly be attributed to an individual, for instance, innate knowledge or knowledge gained through perception, these seem so unimpressive as to not intuitively be credit-worthy at all. If so, then an equivalent challenge might be posed to me: small-scale individual technological activities are not credit-worthy.

My first line of response is that I do not see why credit cannot be something that admits of degree. The possessor of knowledge gained through perception deserves some modicum of credit for being aware of their surroundings and interpreting objects correctly—consider how you might play "Eye Spy" with a toddler and attribute credit to them for their successful participation.

Secondly, I feel that the technological endeavor generally avoids Lackey's objection in the knowledge case, as the vast majority of instances of technological creation are intuitively credit-worthy, with no mental gymnastics required. It just seems obvious that making a home-cooked meal and knitting a scarf deserves some credit. They are things I would complement a friend on. They are activities worth spending time on. Perhaps the closest equivalent to innate knowledge would be a person's channeling of hunter-gatherer technological instinct, for instance, picking up a boulder or stick for use as a weapon. My intuitions in this case are unclear, but either way, this is not the sort of technological situation that is relevant to our present lives (whereas prima facie uncredit-worthy knowledge is still a feature of our lives).

So, in summary, this credit-based account of technological value seems to be promisingly close to common-sense intuitions in grand cases like the production of a spacecraft, and in more local settings like the production of a home-cooked meal. The credit account captures a certain appreciability that societies can have for engineers and their produce, and the wider collective social endeavor that to some extent we all participate in when we make artifacts, big and small. These fundamental components of human life, the creation of the thing as opposed to the thing itself, elevate the value of a technology beyond its value as a mere means.

6.4 Argument 4: Appeal to the Misattributed Value for Scientific Knowledge

It is a much more common and older notion in philosophy that there is intrinsic value in knowledge (or parts thereof). This notion is self-evident in the etymology of "philosophy." Building on the ancients, modern epistemologists have argued for intrinsic epistemic values including for knowledge and for epistemic virtues (see [4, 14, 17, 28])

One potential source of objection to technology being intrinsically valuable is that the value I attribute to technology is in fact just a covert form of epistemic value, for the scientific knowledge underpinning technological endeavors. The realm of achievement that is credit-worthy in the previous section is thereby just epistemic achievement and not some distinct realm called technological achievement.

I suspect the opposite is true. A significant amount of the positive valence toward science is misattributed, when in fact what is valued also is *technological* endeavor, i.e., engineering. Before we can understand such claims, we first need to get clear on the interplay between the two.

Broadly speaking, I advance that scientific endeavor is concerned with the creation of knowledge propositions, while technological endeavor (i.e., engineering) is, as I've defined it, the creation of intentional functional artifacts. Skolimowski [22] puts it in a more artful way: say that "*the scientist is concerned with how things are, but the engineer with how things ought to be.*" Although, a quibble is that we might wonder how engineers get their "oughts" if not from more normative disciplines, e.g., politics and ethics.

Another commonly discussed idea is that technology is "applied science." Can this be true if technological endeavor predates science? I think that, in principle, an applied version of something can exist before its corresponding theoretical version with no issue, the issue is only if the application of science requires an agent's self-conscious knowledge that "they are applying science."

My preferred treatment of the two activities is to see them as lying on a continuum of different processes. Often one is simultaneously engaged in the generation of scientific and engineering propositions as they are building technological concepts. Whether that be synthetic biologists tweaking cells in a lab to understand cell function, an astrophysicist building computer simulations to mirror features of the universe, or a big pharma chemist developing a molecule for a new drug. A popular representation of the intermediate activities between the pure production of scientific knowledge and the pure production of technological artifacts is the widely used "technology readiness level" (TRL) framework developed by NASA [13] but appropriated for use by the European Union [3]. This 9-point scale charts the origin of a commercialized and ubiquitous technology, first in pure science, then applied science, then in prototyping, and then in scaling and mass production.

Consider the following statement of intrinsic value for a particular area of science: synthetic biology. This is from Kaebnick [10], remarking on the work of Ed Regis' book What is Life [15].

> [Synthetic biology] is a morally noble activity, in and of itself. The pursuit of science is the advancement of human mastery, and synthetic biology exemplifies human mastery especially dramatically [...] It is humanism at its most grandiose.

Kaebnick here is articulating what he sees as an intrinsic valuation of science. But this is a misattribution of value to science, when it is in fact for the production of technology. The field of synthetic biology's expressed aims are to extend or modify the behavior of organisms and engineer them to perform new tasks. Synthetic biology is, by definition, a technological discipline, further along the technology readiness level spectrum than pure science. Clearly in this case, mastery consists not just in the accumulation of scientific propositions but an application of them to synthesize, i.e., create, new (semi-)artifacts.

Pure science alone does not confer human mastery of nature. Romantic era scientists like Friedrich Schelling and Sir Humphry Davy made it clear that they only want to learn about nature, not meddle with it. The endeavors that in fact yield mastery are those running all the way along the Technology Readiness Level spectrum. And so, it seems if these are the sort of sentiments that account for some intrinsic value, then all the steps along the TRL should be considered intrinsically valuable.

To see that a reduction of value to mere knowledge is a mistake, take the following case: Suppose a group of NASA rocket scientists and brilliant mathematicians decided to keep all their knowledge propositions to themselves, and though they had collective knowledge of spaceflight, never actually built a spacecraft. In such a case, the propositions are equally well known among NASA constituents. However, this state of affairs seems considerably less valuable.

One might counter that the difference is down to the actual building of the spaceship being a surefire demonstration of the knowledge. But this still falls short. Suppose the scientists simulated the spaceflight on sophisticated computer models, such that we could be very sure the propositions were true. Even then, it still seems less good to me than had they actually done it. Put another way, there are infinitely many technologies that could exist if the relevant knowledge-owners happened to get together and produce them, but the fact that these knowledge items exist already now is not a reason to appreciate the state of affairs very much.

Another example might make the point clearer. Many people appreciate record-breaking technologies. Perennial bestseller "The Guinness Book of World Records" features a page on the world's tallest buildings [9, p.124]. It is currently the Burj Khalifa, which, at over 800 m, is more than five times the height of the Great Pyramids at Giza and twice the height of the Empire State Building. I, along with many other readers, am impressed with the tallest buildings even when their technological functions return me no instrumental value—I might never visit the building but nevertheless be inspired, awestruck, or just merely quite interested in the fact that it has been built.

Many types of knowledge will be involved in the construction of this tall building. But crucially, the propositions that are required to be known to build this tall building are not hugely different to those which contributed to shorter buildings. Moreover, we seem to already know exactly what it would take for us to build *even taller* buildings. And yet, me and other fans of the Guinness World Record books are still impressed by the tallest building as it stands today. In both cases, I reserve at least some appreciation for the actual doing of the activity, the application and employment of propositions, not just the propositions themselves. In this way, technology's value doesn't seem reducible to propositions, even though it could account for some significant portion of my positive valence.

Gilbert Ryle's concept of "know-how" [19] might apply to the manufacture of technology, with impossible-to-articulate "rules of thumb" threatening an account which reduces the knowledge discipline of technology to mere propositions. Engineers in the design and manufacture of spaceships and tall buildings will of course use their know-how along with propositional knowledge. Engineers learn much of their craft on the job and might have skills not articulable in propositional terms. Know-how is indisputably an input into the creation of technology, and it might be epistemically valuable.

But the above examples again show that the existence of some amount of know-how does not sufficiently cover the intuition of technology having value. While I appreciate the know-how of engineers, I appreciate it more that they actually employ this know-how to create remarkable instances of technology like spaceships and tall buildings. We need the referent of our valuation to distinguish between states of affairs where the engineers do actually use their know-how successfully in the creation of these great intentional functional artifacts, and when they don't.

There is no denying the intimate relationship between scientific knowledge (propositions and know-how) and engineering (the creation of intentional functional artifacts). It is worth observing, and marveling, at how incredibly pumped full of knowledge technologies are. At each point of the Technology Readiness Level spectrum, there are distinct fields of scientific knowledge involved. Whether that be progress in every field of fundamental science which allows the generation of fundamental scientific concepts, or moving along the TRL, of engineering and production concepts, or knowledge of gaps in the market, or of the best distribution mechanisms to bring objects to a wide audience.

Technologies act as uniquely rich domains of knowledge; they are substrates about which new sets of propositions can be generated. However, to reduce the value of technology to only the value of the knowledge propositions and know-how would be a mistake, as over and above the knowledge itself is the successful application of the knowledge into producing functional artifacts.

6.5 Technology "Embodying Values"

It is worth noting how the claims of this chapter differ from another discourse in the contemporary philosophy of technology. Some theorists discuss whether technology can *embody* the values of its makers. Akrich [1] introduced the notion of a script, a set of normative expectations and assumptions that are embedded within a technological object. Akrich and others challenge the commonsense "Value Neutrality Thesis": that technology is value neutral, and values are only present in the producers and users. This view is typified by the phrase popularized by the American National Rifle Association that guns don't kill people, people kill people [26].

A much-discussed example posed by Winner [27] is an overpass bridge built by Robert Moses which was intentionally designed to be extraordinarily low for racist purposes: the low overpass would not allow buses through, meaning only those who could afford a car could access the beach on the other side. The thought is that *the bridge itself* embodies the value (disvalue) of racism. De Poel and Kroes give another example of a speed bump, which, for them, embodies a value of human safety. Katz [11] similarly argues that the evil of the Nazi death camps was designed into the technological artifacts themselves. For such philosophers, certain technological objects can have value "baked into them" and so be intrinsically evil or good.

I find the evocative examples in this literature, from speed bumps to death camps, can be equally well explained as holding a very high instrumental (dis)value toward the particular intrinsically (dis)valuable end in question. Rather than being an instance of evil (i.e., an instantiation of an intrinsically disvaluable abstract property), it is equally well explained as an instrument or means to bring about the disvalue. What it seems like they are talking about is the expressive function of artifacts: certain artifacts like bridges and death camps can be means of expressing certain intrinsic values, just as an artwork might express a social critique from a political standpoint.

De Poel and Kroes use the example of a speed bump as having intrinsic value as it embodies the value of human safety. But this misunderstands the point of intrinsic value, that the object is valued for itself. If human safety happened to be better achieved through other means, let's say, because it turned out that the speed bumps did more damage to car's suspension and increased the likelihood of cars malfunctioning, then it is very odd to assert that the speedbump embodies value for human safety. Perhaps the faulty speedbump embodies value for human safety… and negligence? Or does negligence prove there was no true value for safety in the first place?

Crucially, it seems that while these philosophers speak of certain technologies having intrinsic or embodied value, they are not suggesting what I have advanced here, which is that we add the concept of technology itself to the list of concepts we might intrinsically value. Rather, their endeavor is about observing more commonly held and well understood intrinsic values such as for human life (in the case of guns), or racial equality (in the case of the bridge) which they see technological objects as somehow possessing or rejecting. The axiological analysis of these technologies finds nothing special about technology perse, it just reduces to the other intrinsic values that happen to be associated with a cherry-picked selection of particular instances of technology.

To recap this section, here I have offered four strategies for arguing for Technology's *suis generis* intrinsic value. The first was just to appeal to brute intuition—spaceships are cool and there is nothing more to be said about the matter. The second strategy was to explain how technology offers a unique form of liberty, physical liberty, which is analogously valuable to how eminent philosophers like Amartya Sen have argued for other forms of liberty (social, economic) being valuable. The third was an argument that the act of technology production is a virtue, and, moreover, one that is not just attributed to an individual, but rather a profound collective human achievement. The fourth was to recognize that some more widely accepted intrinsic value for scientific knowledge is in fact owed to technology. These arguments are a smorgasbord, aiming to show the potential backing that the intrinsic value of technology camp can accrue.

So, we now have arguments populating either side of the nature-technology political value spectrum. In Part 3 of this book, we will now see our spectrum in action. We will start with an example of how these arguments are made, with force, in a particular area of heated policy debate: the topic of human enhancement biotechnology. We will then see what the spectrum gives us in explanatory power, compared with other theories.

References

1. M. Akrich, The description of technical objects. in *Shaping Technology/Building Society. Studies in Sociotechnical Change*, ed. by W. Bijker, J. Law (MIT Press, 1992), pp. 205–224
2. I. Berlin, *Two Concepts of Liberty* (Clarendon Press, 1958)
3. European Union, *Technology Readiness Levels (TRL)* (Extract from Part 19—Commission Decision C(2014)4995) (2014)
4. J. Greco, in *Knowledge as Credit for True Belief*, ed. by V. DePaul, L. Zagzebski (2003), pp. 111–134
5. J. Greco, The nature of ability and the purpose of knowledge. Philosophical Issues **17**, 57–69 (2007). https://doi.org/10.1111/j.1533-6077.2007.00122.x
6. J. Greco, Knowledge and success from ability. Philosophical Stud. **142**(1), 17–26 (2009). https://doi.org/10.1007/s11098-008-9307-0
7. J. Greco, A (different) virtue epistemology. Philosophy Phenomenol. Res. **85**(1), 1–26 (2012). https://doi.org/10.1007/s11098-008-9307-0
8. J. Greco, in *The Transmission of Knowledge* (Cambridge University Press, 2020).https://doi.org/10.1017/9781108560818

9. Guinness World Records, Record breaking buildings. in *Guinness World Records* (2016), pp. 124
10. G. E. Kaebnick, Engineered microbes in industry and science: a new human relationship to nature? in *Synthetic Biology and Morality (Basic Bioethics)* (MIT Press, 2013), pp. 51
11. E. Katz, On the neutrality of technology: the Holocaust death camps as a counter-example. J. Genocide Res. **7**(3), 409–421 (2005). https://doi.org/10.1080/14623520500190363
12. J. Lackey, Why we don't deserve credit for everything we know. Synthese **158**(3), 345–361 (2007). https://doi.org/10.1007/s11229-006-9044-x
13. NASA, *Technology Readiness Levels* (2012). https://www.nasa.gov/directorates/heo/scan/engineering/technology/technology_readiness_level
14. D. Pritchard, J. Turri, J. A. Carter, The value of knowledge. in *The Stanford Encyclopedia of Philosophy*, ed. by E. N. Zalta, U. Nodelman (Fall 2025 ed., 2025). https://plato.stanford.edu/archives/fall2025/entries/knowledge-value/
15. E. Regis, in *What is Life?* Farrar, Straus, and Giroux (2008)
16. W.D. Riggs, What are the 'chances' of being justified? Monist **81**(3), 452–472 (1998). https://doi.org/10.5840/monist199881319
17. W.D. Riggs, Reliability and the value of knowledge. Philosophy and Phenomenol. Res. **64**(1), 79–96 (2002). https://doi.org/10.1111/j.1933-1592.2002.tb00143.x
18. W.D. Riggs, Two problems of easy credit. Synthese **169**(1), 201–216 (2009). https://doi.org/10.1007/s11229-008-9342-6
19. G. Ryle, in *The Concept of Mind* (Hutchinson, 1949)
20. A. Sen, *Commodities and Capabilities* (Oxford University Press, 1985)
21. A. Sen, Markets and freedoms: achievements and limitations of the market mechanism in promoting individual freedoms. Oxford Econ. Papers **45**(4), 519–541 (1993)
22. H. Skolimowski, The structure of thinking in technology. Technol. Culture **7**(3), 371–383 (1966). https://doi.org/10.2307/3101935
23. E. Sosa, The raft and the pyramid: coherence versus foundations in the theory of knowledge. Midwest Stud. Philosophy **5**, 3–25 (1980)
24. E. Sosa, in *The Place of Truth in Epistemology*, ed. by V. DePaul, L. Zagzebski (chap. 7) (2003)
25. E. Sosa, in *Apt Belief and Reflective Knowledge* (Vol. 1): *A Virtue Epistemology* (Oxford University Press, 2007). https://doi.org/10.1093/acprof:oso/9780199297023.001.0001
26. I. Van De Poel, P. Kroes, Can technology embody values?, in *The Moral Status of Technical Artefacts*. ed. by P. Kroes, P.-P. Verbeek (Springer, 2014), pp.103–124
27. L. Winner, Do artifacts have politics? Daedalus **109**, 121–136 (1980)
28. L. Zagzebski,in *Intellectual Motivation and the Good of Truth*, ed. by V. DePaul, L. Zagzebski (2003), pp. 135–154

Part III
The Nature-Technology Political Value Spectrum in Action

Chapter 7
A Case Study—The Human Enhancement Debate

Abstract This chapter employs the Nature-Technology intrinsic value spectrum as a metacommentary to reframe the philosophical debate over the unnaturalness of human enhancement technologies (synthetic biology, genetic engineering, etc.). The chapter examines three camps in the naturalness debate: hard-line opponents (e.g., Leon Kass, Michael Sandel, Francis Fukuyama) who argue that tampering with human nature is intrinsically wrong; staunch proponents (e.g., transhumanists) who reject the intrinsic value of naturalness; and moderates who seek to draw a reasoned line in the sand between acceptable and unacceptable alterations. The central argument is that this *prima facie* ethical debate, which seeks an objective, universally applicable cut-off point, constitutes a category mistake. By reducing the conflict to a specific instance of the broader Nature-Technology intrinsic value trade-off, the chapter concludes that the debate is fundamentally a political problem of balancing antagonistic intrinsic values, not an ethical problem of finding binary rights and wrongs. This political lens alleviates the pressure on bioethicists to find a "holy grail method" for drawing lines, demonstrating the spectrum's utility in diagnosing and managing complex, value-laden public policy conflicts.

Keywords Human enhancement ethics · Bioethics · Synthetic biology · Transhumanism · Naturalness argument · Genetic engineering

In this chapter, I aim to show how the Nature-Technology political spectrum can be employed to shed light on an area of lively debate. For this deep dive, I have chosen debates concerning the unnaturalness of human enhancement technologies, but I could have equally chosen raging debates in the application of artificial intelligence, environmental policy, or many others.

Our ability to manipulate the fundamental components of the human body is rapidly changing with the advancement of the loose collection of disciplines termed "synthetic biology." Synthetic Biology combines biology, engineering, and related disciplines to design and chemically synthesize DNA to create organisms, including human beings, with novel or enhanced characteristics and traits.

B. Steyn, *Technology vs Nature*,
SpringerBriefs in Applied Sciences and Technology,
https://doi.org/10.1007/978-3-032-18582-2_7

Bioethicists commonly distinguish two potential rationales for altering humans. A relatively non-controversial use is therapeutic, where the objective is to bring a person deficient in some characteristic up to "normal functioning" (if such a thing can be defined). More controversial is the idea that we might actively innovate to enhance humans beyond normal functioning. Paradigmatic examples of non-therapeutic human enhancement include biosynthetic growth hormone for increased stature; athletic doping to increase endurance and strength; psychopharmaceutical interventions for improving memory, mood, and cognitive capacity; and a range of profound genetic and neurological interventions to increase lifespan, which bring new sensory-motor abilities and alter our personalities to encourage more empathy and generosity.

These technologies are controversial with respect to many axes of value. For instance, there is a burgeoning public and philosophical debate around the perceived unfairness of athletic doping and similar implications on use of artificial enhancements in society at large. There are also lots of ethical concerns around the proper use and explorations of these technologies and the risks and harms they pose.

My focus in this chapter is on one particular debate upon which there has been an abundance of bioethics literature. This is the debate focused on the claim that because human genetic engineering is unnatural, it therefore is intrinsically wrong. I will examine in turn three camps in this debate: There are those who believe that all human-centered genetic engineering is intrinsically wrong because it is unnatural. There are those who support human genetic engineering as either intrinsically right, or not intrinsically wrong (and therefore ought to be encouraged if the instrumental benefits are shown to be high). There are those in-between, searching for a way to delineate some limits at particular technological junctures or with regard to particular human traits.

7.1 The Naturalness-Based Attack on Human Enhancement

Many feel strongly that human nature is something that ought to be protected from the technological impulse. A leading voice on the side against human enhancement technology, and a former Chair of the President's Committee on Bioethics, is Leon Kass. Kass has argued in multiple places that our human bodily constraints, especially the fact of human finitude, must be maintained for human flourishing [10, 11]. Regarding human cloning technologies, which can be construed as an enhancement or alteration to human's natural reproductive systems, Kass argues that cloning personifies our desire fully to control the future, while being subject to no controls ourselves. *"Enchanted and enslaved by the glamour of technology, we have lost our awe and wonder before the deep mysteries of nature and of life"* [11, p. 683]. Kass appeals to the wisdom of repugnance, arguing that peoples' feeling of disgust toward cloning and other human biotechnologies ought to be treated as an expression of a deep and inarticulable wisdom that it is morally wrong.

Sandel [18] takes umbrage with our yielding to the human drive to mastery, arguing that this may destroy our appreciation of the gifted character of human powers and achievements. Meanwhile, George Annas, advancing the need for a new U.N treaty to ban species-altering research, argues inheritable genetic alterations and cloning can be seen as crimes against humanity, in altering the essence of humanity itself. Similarly, Francis Fukuyama believes in some universal human essence he labels "Factor X." He is concerned that human enhancement may violate Factor X, challenging biological human nature as the appropriate basis of shared human rights and dignity, and so eroding our sense of common humanity [5]. Those using technology to manipulate natural beings, especially humans, are often accused of playing god. But the emotive force behind the metaphor of playing god does not depend on a theistic worldview [15].

On the other hand, proponents of human enhancement, particularly those comprising the transhumanism movement, commonly reject the intrinsic significance given to human nature. They see the above arguments as an unsubstantiated, bald assertion that human enhancement is wrong because it is unnatural. But, as Streiffer [19] suggests, this provides no argument to those who do not already share the view that unnaturalness equals wrongness. Mandle [13] notes that, contrary to Kass, the public are not universally repulsed by human-centered biotechnologies like cloning. Preston's [17] response to Kass's wisdom of repugnance argument is that we just have to get over it. Harris [9], in arguing we are morally obliged to innovate on human biotechnology, contests the preoccupation with nature, noting that famines, floods, droughts, and storms are all natural and all disastrous. Contending that awe and mystery is necessarily wherein value lies, Caplan [3] notes that the value of life is not cheapened by understanding how it works, but rather, like a fine wine or jazz music, it can enhance appreciation.

McConnell [14] takes aim at arguments given by Fukuyama and Annas, pointing out that much of these authors' negative valence to post-humanity is left unexplained. He challenges us to consider whether losing one's humanity, as Fukuyama puts it, is such a terrible outcome, particularly when set against the potential health and wellbeing benefits of human genetic engineering technologies. He contends that the risk of reductionism of ends Fukuyama sees is but another instrumental risk and not an intrinsic objection. He also challenges Annas to clarify exactly why creating a new species of post-humans is in and of itself a bad thing.

Glover [6] defends a similar sentiment in the use of biotechnologies in the specific case of producing designer babies. He argues: *If a good argument showed that some terrible characteristic—which by genetic means, we could change—was essential to being human, it might be better to transcend the limits of humanity rather than stay as we are. The idea of what is essential is a murky one, but, even if it were not, its importance is unclear* [6, p. 84]. Bostrom [2] answer to this murkiness is to claim that in fact there are no static features of humanity at all. Our moral kinship and forms of social organization have changed radically over time, and it is difficult in his view to identify anything we might want to call an essentially human vulnerability.

For some, the playing god metaphor ought to be abandoned, not saved from its archaic religious origins. Kaebnick [7] documents Drew Endy, a leading figure in the field of synthetic biology, saying that the questions of playing God or not are so superficial and embarrassingly simple that they're not going to be useful.

We can see the intrinsic value sentiment in favor of human enhancement echoed strongly among the Transhumanist movement. Anthropologist Abou Farman's *On Not Dying* [4] documents those groups seeking immortality. Their faith in technology, he argues, is quasi-religious. It is not merely that living longer provides some instrumental benefit, it is that conquering death is, for this group, a triumph of man's ingenuity over nature. Newton Lee, author of *The Transhumanism Handbook* [12], argues with a positive valence that the fusion of man and machine is inevitable. One notable proponent of the Transhumanist movement is billionaire co-founder of Paypal, Peter Thiel. An interviewer of Thiel writes extensively about his transhumanism. "Thiel [thinks that] the problem with transhumanism isn't that it seeks to remake humanity, but that it isn't ambitious enough in this regard" [8, para.20].

At this stage in the dialectic, some commentators search for some reasoned way to distinguish between those sorts of tampering with human nature that are acceptable and those that aren't.

Christopher Preston's paper "Drawing lines in Darwin's Sand" [16] attempts to distinguish between the more traditional practices of molecular biology, which involve tweaking and modifying existing organisms with new genes, and the more advanced areas of synthetic biology which involve creating an entirely new organism. The latter, and only the latter in Preston's view, crosses the line. The former, he argues, cannot be objected to as it is how Darwinian evolution works. The latter, meanwhile, is entirely new. Tweaking of the human genome then, would on this account be an acceptable form of biotechnology.

However, Bedau and Larson [1] challenge this distinction twice-over, firstly as a series of tweaks can result in quite profound changes. There is a Sorites-Paradox-esque continuum between the two types of endeavor. Secondly, even if there is some category distinction between novel synthetic organisms and genetic modification, it's not clear why the violation of category boundary is morally relevant. It seems then that, while the extremes of total endorsement of human nature and total permissiveness to enhancement technology are often seen as undesirable among bioethicists, no definitive line in the sand has been successfully drawn.

7.2 Reduction to the Nature-Technology Value Spectrum

I do not aim to wade in on this debate, but rather to offer a metacommentary. What should be clear at this stage of the book is that the debate over human enhancement's unnaturalness is simply not interesting. Well, not philosophically interesting. Because it is just one instance of the wider values trade-off we have outlined in previous chapters, concerning the way humans interact with not just other humans, but *all* physical material.

Those who object to human enhancement we can wed with those discussed in earlier who generally intrinsically value nature. The positions documented above are extremely like how environmental ethicists write on preserving other natural entities. Recall Elliot's example of the restored natural habitat being akin to art forgery. He values the history, authenticity, awe, and majesty of nature as being untouched by humans. It is these same features invoked here to justify preservation of human nature. For instance, take Michael Sandel's calling enhancement desires hubristic (2004), he is directly opposed to the idea that technological mastery of nature is a virtue. Rather than trying to strive for technological progress in relation to our bodies, he wants us to appreciate our authentic, natural bodies.

The notion of "Playing god" is invoked not just in discussion of human genetic tampering, but more widely, for instance, in debates around ways in which technologies interfere with the natural reproductive process, including sex changes, abortion, IVF treatments, embryo research, and contraception. But also, with regard to humanity's subduing of natural habitats. For instance, climate geoengineering. We've also already seen that the same metaphor has been used with regard to other types of synthetic biology beyond human life.

Meanwhile, we can wed the transhumanists and other pro-enhancement theorists with the intrinsic value of technology positions discussed earlier. This debate mirrors the wider discourse in often ignoring the potential for intrinsic value from technological endeavors. But with a charitable interpretation of Kaebernick's analysis, we can see the positive "intrinsic value for technology" view is represented. Instead, the defensive case focuses more on undermining appeals to the intrinsic value of nature.

Another approach to showing that this is in fact one instance of the general debate is to see that all the positions documented in previous chapters can be straightforwardly applied to the subject matter of human beings, even if that was not the authors' intention. What was required to apply these views was for human beings to have a metaphysical status of semi-artifact, which we provided a strong argument for earlier. Just because those general arguments weren't intended to apply in this specific case, we have no reason to think that they are any less potent.

Let us consider some potential objections to my reduction of the human enhancement naturalness debate to the nature/technology value spectrum. Firstly, it is logically possible that one might take one stance concerning human nature, and another concerning nature of other sorts (animal, plant, habitat, process). While it would seem curiously arbitrary to intrinsically value all of nature except human nature. One plausibly non-arbitrary reason for prioritizing human nature and not nature in general might be a metaethical commitment that ethics is about what is good *for* humans, and so it is specifically an account of human nature that matters to provide a basis for a coherent ethics, but not necessarily nature in general.

However, it is not easy to de-couple human nature from nature in general. Human nature is what it is because humans have evolved to adapt to particular external environments. All our physical and mental features would have been otherwise if the environment were otherwise. Even if it was only a conception of human nature required to buttress one's ethics, it seems that this human nature would not necessarily continue to entail the good for humans when there is a substantial shift in the context to which that human nature was fitting to. Anti-enhancement theorists seek us to retain awe, wonder, and acceptance for both the gifts and limitations of humanity. But what we consider gifts and limitations of human bodies are context dependent. Given human nature must be conceived within a broader context, those intrinsically valuing human nature cannot help but also support, as a means to its creation, the wider states of affairs within which human nature resides in nature in general.

Even if we could successfully de-couple human nature value considerations from nature in general, we still face the brute metaphysical fact that the former is a subset of the latter, and so human enhancement debates face the same general dialectic covered in earlier chapters.

A second challenge which, *prima facie,* threatens to break the dichotomy of (human) nature versus (human-enhancing) technology is the line of argument raised by Jose Ortega y Gasset, that *"man without technology is not man."* As addressed in Chapter 4, the counter-argument that the technological impulse is "natural" is problematic, as human nature contains contradictory impulses (e.g., to preserve versus to innovate); merely asserting a natural origin fails to provide a *justification* for which impulse should dominate.

But furthermore, the appeal to the naturalness of human innovation is also potentially self-defeating. If there is a human impulse to innovate, and this innovation can be applied to humans themselves, and this impulse ought to be respected to achieve betterment in the species, then it appears we could, in the eyes of some, better our species by *removing or curbing* our impulse to innovate in the first place! Again, we need to find independent reasons to justify which impulses to curb and which to honor.

7.3 Not Interesting?

Why exactly is it not interesting to consider a single instance of a wider debate? I don't doubt that in many respects it is interesting to follow through the implications of the general stance to different areas. It is also true that focusing on the specific instance can helpfully illustrate and reveal general concerns, and indeed that is why our focusing down here is a worthwhile endeavor as part of my overall project. But what I mean is that the specific instance does not generate many of its own considerations or dimensions over-and-above the general debate.

Let's consider an analogy elsewhere. There is a rich and complex debate around the merits and demerits of redistributive justice. We can have such debates without considering in detail the specific mechanisms of redistribution, for instance, whether we use income tax, capital gains tax, and corporation tax. When debating some particular tax, it is likely that part of the core rationale for it links to a general stance on redistributive justice. The general stance is sufficient. We don't have a "philosophy of income tax," and if we did, it would not be particularly interesting beyond the general debate on redistributive justice. Similarly here, we don't need a specific branch of bioethics dealing with how our general spectrum pertains to human synthetic biology; we just need the nature/technology spectrum.

Viewing this issue at a political level of explanation helps to meet the challenge of drawing lines in the sand. We shouldn't expect an answer in the form of precise technology-specific statements like "*Recombinant DNA techniques are wrong when they edit a particular individual's traits beyond the population's range of naturally occurring variation.*" On the political conception, drawing such delineations would be analogous to a socialist attempting to identify the single, numerically correct tax rate that perfectly delineates "fair" from "unfair." The socialist does not face such an absurdity because the tax debate is carried out at the more general, dispositional level of whether to value equality or liberty. Similarly, once we see the human enhancement debate in the terms of the nature-technology political spectrum, we should recognize that the endeavor to argue for some correct, universally applicable cut-off point for a particular technological intervention is hopeless; it is a misplaced ethical demand on a fundamentally political problem.

In this chapter, we have explored the heated debate around the naturalness of human-centered synthetic biology. We have seen those on the pro-nature side wanting to protect something special about natural humanity. We have seen those on the pro-technology side wanting to "play god" in pushing the species boundary, and we have seen those struggling in the middle to draw a coherent line in the sand.

By re-casting this debate in terms of my more general nature/technology value spectrum, this is perhaps an unsatisfying position. I have not been unequivocal in whether particular enhancements are right or wrong, and that seems to be what many bioethicists want to know. Unfortunately, those asking this question are making a category mistake. As with other political dispositions, we should not expect them to create binary rights and wrongs out of particular policies, rather, they influence at a more general level.

These debates are also pertinent in the field of artificial intelligence. People want to draw the "Darwinian line" in choosing what applications of AI are acceptable and not acceptable with respect to preserving a distinct human cognitive space. Unfortunately, for the same reasons discussed here, that is not a sensible ethical debate and must be construed as a political one.

Now that we have the theory firmly on the table, it is time to examine it in its entirety and ask: is this an original contribution, first in philosophy and then in political science?

References

1. M. Bedau, B. Larson, Lessons from environmental ethics about the intrinsic value of synthetic life. in *Synthetic Biology and Morality: Artificial Life and the Bounds of Nature*, ed. by G. E. Kaebnick (MIT Press, 2013)
2. N. Bostrom, Existential risks: Analyzing human extinction scenarios and related hazards. J. Evolut. Technol. **9** (2002)

3. A. Caplan, in *Now Ain't that Special? The Implications of Creating the First Synthetic Bacteria* (Scientific American, 2010). https://blogs.scientificamerican.com/guest-blog/2010/05/20/now-aint-that-special-the-implications-of-creating-the-first-synthetic-bacteria/
4. A. Farman, in *On Not Dying: Secular Immortality in the Age of Technoscience* (University of Minnesota Press, 2020). https://doi.org/10.5749/j.ctv10rrc6z
5. F. Fukuyama, *Our Posthuman Future* (Farrar, 2002)
6. J. Glover, in *Choosing Children: The Ethical Dilemmas of Genetic Intervention* (Oxford University Press, 2006)
7. G.E. Kaebnick, Engineered Microbes in Industry and Science: A New Human Relationship to Nature? *Synthetic Biology and Morality* (Basic Bioethics) (MIT Press, 2013), pp. 51
8. M. Harrington, in *Peter Thiel on the Dangers of Progress: The Tech Billionaire Discusses Silicon Valley, Christianity and Apocalypse*. Unheard (2022). https://unherd.com/2022/12/peter-thiel-on-the-dangers-of-progress-2/
9. J. Harris, Enhancements are a moral obligation. in *Human Enhancement*, ed. by J. Savulescu, N. Bostrom (Oxford University Press, 2009), pp. 134
10. L. Kass, *Beyond Therapy: Biotechnology and the Pursuit of Human Improvement*. Paper Prepared for the President's Council on Bioethics (2003)
11. L.R. Kass, The wisdom of repugnance: why we should ban the cloning of humans. Valparaiso Univer. Law Rev. **32**(2), 679–705 (1998)
12. N. Lee, *The Transhumanism Handbook* (Springer, 2019)
13. J. Mandle, Synthetic biology and public reason. in *Synthetic Biology and Morality: Artificial Life and the Bounds of Nature*, ed. by G. E. Kaebnick (MIT Press, 2013), pp. 131
14. T. McConnell, Genetic enhancement, human nature, and rights. J. Med. Philos. **35**, 415–428 (2010). https://doi.org/10.1093/jmp/jhq034
15. G. McKenny, Nature as given, nature as guide, nature as natural kinds: return to nature in the ethics of human biotechnology. in *Without Nature? A New Condition for Theology*, ed. by D. Albertons, C. King (Fordham University Press, 2009), pp.152–177
16. C.J. Preston, Synthetic biology: drawing a line in Darwin's sand. Environ. Values **17**(1), 23–39 (2008)
17. C. J. Preston, Synthetic bacteria, natural processes, and intrinsic value. in *Synthetic Biology and Morality: Artificial Life and the Bounds of Nature*, ed. by G. E. Kaebnick (MIT Press, 2013), pp. 107
18. M.J. Sandel, The case against perfection: what's wrong with designer children, bionic athletes, and genetic engineering. Atl. Mon. **292**(3), 50–62 (2004)
19. R. Streiffer, Human/non-human chimeras. in *The Stanford Encyclopedia of Philosophy*, ed. by E. N. Zalta (Summer 2019 ed., 2019). https://plato.stanford.edu/archives/sum2019/entries/chimeras/

Chapter 8
The Nature-Technology Political Spectrum Versus Philosophy

Abstract This chapter defends the methodological and conceptual novelty of the Nature-Technology Political Spectrum by positioning it within contemporary philosophy. It directly addresses the problem of disciplinary siloing, citing David Kaplan's observation that environmental ethics and the philosophy of technology remain isolated, arguing that the Nature-Technology spectrum serves as a successful application of Occam's Razor by identifying a simple, unifying political thread across disparate debates. Developed within the foundational assumptions of analytic philosophy (prioritizing conceptual clarity and descriptive reduction), the framework deliberately contrasts with Continental philosophy, which is often characterized by "world-building," normative claims, and a rejection of descriptive scientific reductionism. Despite this fundamental methodological divergence, the framework is shown to resonate with critical Continental observations, particularly from thinkers in New Materialism and the "humanities philosophy of technology" (e.g., Donna Haraway, Michel Serres) and Nick Land's Accelerationism. The chapter details how the spectrum unifies disparate fields, while maintaining a modular separation from economic and social political spectra—the final goal being a set of unbiased, non-normative political coordinates for empirical testing.

Keywords Analytic versus continental philosophy · Philosophy of technology · Environmental ethics · Occam's Razor · Descriptive reductionism · Nature-artifact continuum · Donna Haraway · Bruno Latour · Jacques Ellul · Accelerationism · Nick land

As we have seen throughout this book, there is extensive scholarship covering the constituent parts of this theory. The definitional components, intrinsic value axiology, nature, and technology, find deep scholarship in different philosophical subdisciplines. We have a deep discourse in environmental ethics and bioethics around the intrinsic value of nature. Advancing the positive intrinsic value case for the specific form of technology relating to human bioenhancement, we have the transhumanists. We also have some historical understanding of this same tension given by historical scholarship on the enlightenment and romanticism.

B. Steyn, *Technology vs Nature*,
SpringerBriefs in Applied Sciences and Technology,
https://doi.org/10.1007/978-3-032-18582-2_8

But environmental ethics and philosophy of technology remain in silo in the analytic tradition. Consider a passage from David Kaplan in his introduction to a compendium of essays on the subject of philosophy of technology and the environment [6]:

> Technology and the environment are like two sides of the same coin: Each is fully understood only in relation to the other. Yet, despite the ample overlap of questions concerning technology and the environment, the two philosophical fields have developed in relative isolation from each other. Even when philosophers in each field address themselves to similar concerns, the research tends to be parallel rather than intersecting, and the literatures remain foreign to one another.

Directly answering Kaplan's call, my key contribution is in bringing this isolated scholarship together in one framework. But as well as being unifying, the spectrum is also a successful application of Occam's Razor, recognizing that a simple common thread underlies debates in *prima facie* disparate fields. Arguing directly along the nature-technology political spectrum is like submitting a class-action lawsuit to consider all the individual cases in one go. We can reduce esoteric areas of debate into more standardized argumentative moves.

The central arguments in my book are developed exclusively within the foundational assumptions and techniques of analytic philosophy and political science, prioritizing conceptual clarity, logical structure, and semantic precision. The aim is to make a tightly defined, neat claim: namely, that nature and technology occupy a political value spectrum. A claim which can be empirically tested in the social sciences, and operationalized in lay political discourse, for instance, in policy discussions among parliamentarians and civil servants. This methodological commitment necessitates a deliberate exclusion of several prominent continental thinkers whose work touches upon similar themes but is founded on fundamentally different philosophical traditions, which aren't aiming at these objectives. Extensive interpretation and engagement of this body of work would require specialized training in the meta-theoretical commitments of phenomenology, structuralism, and post-humanist critique which is not the remit of this book.

Despite this fundamental methodological difference, my argument finds points of resonance with some critical observations made within the Continental tradition. Namely, the New Materialists, Science and Technology Studies scholars, and what is referred to as the "humanities philosophy of technology." This includes thinkers like Bruno Latour, Michel Serres, Bernard Stiegler, Donna Haraway, Jane Bennett, Jacques Ellul, Lewis Mumford, and Nick Land.

8.1 Breaking Binaries

My argument endorses an axiological spectrum, underpinned by a metaphysical continuum between nature and artifact. The notion that these labels of natural and artificial apply fluidly, not anchored to any particular metaphysical state of affairs, is reminiscent of the position famously taken by Donna Haraway in *A Cyborg Manifesto* [4]. Haraway introduces the notion of a cyborg to challenge the socially constructed boundaries of, among other binaries, human and machine. In the more recent *Staying with the Trouble* [5] Haraway continues by arguing there is no escape from our entanglement with industrialization and capitalism, and we must, as the title suggests, learn to stay with the trouble.

Michel Serres's *Hominescence* [8] argues that humanity is undergoing a profound, epochal transformation, which he calls "hominescence." The book is a philosophical meditation on the radical changes humanity has experienced, particularly over the preceding 50 years, with the human species no longer simply *inheriting* its condition, but beginning to build it itself through technical means, creating objects (like the Internet or the atomic bomb) that are globally impactful, but that humanity does not yet fully control.

Building on Haraway and Serres, and echoes of this idea in the work of Gasset and Stiegler as discussed earlier, my apparatus of a nature/artifact continuum allows us to challenge a strict boundary in the relative natural/artificial content of humans themselves. However, in a divergence from these thinkers, my apparatus allows us to be more precise about changes in artificiality—namely that there is a spectrum, artificiality admits of degree. Such that we can sensibly engage in debates about whether and to what extent we become cyborgs versus retaining our humanity.

Somewhat relatedly, Latour's *The Politics of Nature* [7] critiques an apparent distinction drawn between nature (defined differently here to mean "unquestionable facts about reality," as in "natural laws") and Society. He rejects the view that scientists can discover objective facts about nature and impose them on politics. Latour proposes replacing the nature/society dichotomy with a new collective—a political community that explicitly incorporates both humans and nonhumans (such as microbes, rivers, and climate systems). While ostensibly writing about similar phenomena, Latour's Parliament of Things, in incorporating more than just the human being, is better seen as a theory in the Kantian inherent worth strand of environmental ethics, and therefore not directly engaging with the content of this book.

8.2 The Value-Ladenness of Technology

The value-ladenness of technology has been discussed by thinkers including Martin Heidegger, Lewis Mumford, and Jacques Ellul, and more recently, Don Ihde, Kyle Whyte, Ryan Gunderson, and Brett Clark, in what is described as the "humanities philosophy of technology," where the key line of inquiry concerns technology as a social and cultural force. Here, in characterizing society's woes, the view of technology taken tends to be either that it is intrinsically evil, or that it is a neutral midwife of the evils of mankind.

Most famously, in The Technological Society [3], Ellul argues that society has come to value "instrumental efficiency" as an end-in-itself, and this is a fatal misstep which has dehumanized and destroyed meaning in our lives. The relentless drive toward efficiency in all areas of life (including manufacture of objects) is a mentality that makes us fail to observe human freedom, truth, beauty, and justice. Ellul is not so much concerned with the value/disvalue of technological artifacts per se, but rather humanity's "instrumentalist attitude" to all things. Even so, we could broadly interpret him as falling neatly into the "pro-nature/anti-technology" side of my political value spectrum.

Meanwhile, the idea that technology has *positive* value has seen less explicit attention. But two figures in continental philosophy of technology stand out for bucking the trend. Friedrich Dessauer, German physicist-turned-philosopher, attempted to extend a Kantian philosophical framework by adding a fourth critique: a critique of technological making. While Kant argued that scientific knowledge is limited to the world of appearances, [2] suggests that technological invention allows for direct contact with a transcendent realm of pre-established solutions to technical problems. For Dessauer, the pursuit of technology has the character of a Kantian categorical imperative. Human participation in creation is a religious experience; "the greatest earthly experience of mortals" [2, p. 66].

Mario [1, p. 68] rejects the "romantic wailings" of Ellul and Heidegger, arguing for a more extensive application of engineering and scientific modes of thought into human social affairs. In a sense, he is arguing that technology is good, but, like other continental theorists, he is not so much interested in technological artifacts per se, but about a broader conception of engineering which includes social engineering (e.g., applied sociology, education, administration), and scientific theory creation (e.g., systems theory).

Another relevant school of thought is Nick Land's Accelerationism (2012), which, while originating in the UK at Warwick's Cybernetic Culture Research Unit in the 1980s, is stylistically continental (or at least is *not* analytical). Accelerationism is a radical, nihilistic, and intentionally anti-humanist position, advocating for the speeding up and intensification of technological and capitalist processes, with the objective being a post-human Singularity. Land views capitalism itself as an autonomous self-replicating meme which will inevitably break down all human structures, traditions, and identities. And furthermore, his normative claim is that we should embrace this process. We humans are merely the "meatware" through which this autonomous capitalist force is realizing itself.

Land, in my framework, might be considered someone with an intrinsic value for technological progress. But his work does not neatly reduce down as such, because he is also wedded to other normative commitments, such as the rejection of the importance of other human values relative to this technological imperative, and his work rests on complex empirical predictions around the inevitability of capitalist takeover, and epistemological claims about capitalism having an agency unto itself aside from humans. Some of these claims are modular. Land is a *right* accelerationist, but *left* Accelerationists like [9] believe technology is not inherently an alien force, but a tool that can be *repurposed* once it is freed from the constraint of the profit motive. They see the technological imperative for fully automating production and labor. For them, the productivity gains created by machines should be used to abolish work as the central organizing principle of human life, rather than merely causing mass unemployment and precarity under capitalism.

The fact that there are both left and right accelerationists is further evidence that we can treat the economic and the technological as separate political spectra. That said, these two ideologies also show that it is possible to weave complex arguments which span these two spectra simultaneously.

Despite these resonant points about the nature/technology boundary and about the intrinsic (dis)value of technology, there is a fundamental incompatibility between my work and the Continental philosophy mentioned. First, because these thinkers are ontological world-builders who, through their distinctive literary styles, each make multiple normative claims at once, along with empirical claims, for instance, about the inevitability of technological progress. Meanwhile, I seek only to present one dimension of a multidimensional political life. And upon that dimension, I am merely aiming to describe a political landscape while making no first-order normative claims of my own, nor any empirical claims about, for instance, the rate of technological advancement.

Secondly, many continental philosophers would reject my attempt at a merely descriptive scientific reduction. First, because political reality is too complex to be grasped through the granular, piecemeal form of conceptual analysis favored by the analytic tradition and requires the richer expressive potential of a more literary philosophical tradition. But also, they might critique my supposed lack of normativity. For the accelerationists, their empirical assumptions necessitate a stance on multiple spectra at once, making a clean disaggregation not politically viable. Meanwhile, for many continental and postmodern philosophers of science, perhaps my attempts at scientific reduction are themselves a reflection of some underlying normativity.

I of course do not subscribe to this view. I think what I am doing is politically neutral, aimed at providing clarity on a political landscape, with which one can appraise certain normative positions. My framework allows for a clean modular separation of economic, social, and technological goals. I give unbiased political coordinates which cut through the complexity of non-analytic arguments. And from this framework, we can derive testable predictions through the ordinary language surveys and interview-based methodologies of political scientists, to which I will turn to in the next section.

References

1. M. Bunge, The five buds of technophilosophy. Technol. Soc. **1**(1), 67–74 (1979). https://doi.org/10.1016/0160-791X(79)90009-5
2. F. Dessauer, *Philosophie der technik*. (F. Cohen, 1927)
3. J. Ellul, *The Technological Society* (J. Wilkinson, Trans.). (Vintage Books, 1964). (Original work published 1954)
4. D. Haraway, A cyborg manifesto: science, technology, and socialist-feminism in the late twentieth century, in *Simians, Cyborgs and Women: The Reinvention of Nature*. Routledge (1991), pp. 149–181

5. D. J. Haraway, *Staying with the Trouble.* (Duke University Press, 2016)
6. D. Kaplan, *Philosophy, Technology, and the Environment* (MIT Press, 2017)
7. B. Latour, *Politics of Nature: How to Bring the Sciences into Democracy* (C. Porter, Trans.). (Harvard University Press, 2004)
8. M. Serres, *Hominescence.* (Le Pommier, 2001)
9. N. Srnicek, A. Williams, *Inventing the future: Postcapitalism and a world without work* (2015)

Chapter 9
The Nature-Technology Political Spectrum Versus Political Science

Abstract This chapter explores the inadequacy of existing political science methodologies to empirically validate the theoretical claims of the Nature-Technology Political Spectrum. The central critique is leveled at major political attitude surveys (e.g., World Values Survey, 2-MEV scale) for their consistent failure to disaggregate intrinsic value from instrumental/welfare-based measures. Specifically, the popular 2-MEV scale is scrutinized for two key methodological flaws: (1) its "Preservation" and "Utilization" dimensions are incorrectly found to be orthogonal, a result of not forcing respondents to make choices between mutually exclusive goods; and (2) One-Sidedness, as existing scales fail to posit a substantive, non-ambivalent counter-value to nature. To rectify this shortfall, the chapter concludes by outlining a program for future empirical research, detailing three mandatory criteria for a new survey instrument: explicitly targeting intrinsic value, mandating forced-choice trade-offs, and focusing on the spectrum's core thematic poles (Naturalness: awe, giftedness; Technology: mastery, ingenuity). This new design is required to robustly demonstrate the Nature-Technology spectrum's explanatory power and orthogonality from the established Economic Left/Right and Social Libertarian/Authoritarian spectra.

Keywords Political science methodology · Political attitudes surveys · Value survey techniques · Intrinsic versus instrumental value · 2-MEV scale · Survey design · Post-materialism · Political values research · Nature public attitudes surveys · Technology public attitudes surveys

9.1 Introduction

Political attitude surveys are a common technique in political science. With a sufficiently well-designed question-set, an attitudes survey can be a robust way to make inferences about people's political values, even among those who may not have self-diagnosed or are not aware of the terminology of political ideology. These surveying methods were developed by [12, 15–17], and many others since (see [18]), who have all used value survey techniques in combination with the statistical technique

B. Steyn, *Technology vs Nature*,
SpringerBriefs in Applied Sciences and Technology,
https://doi.org/10.1007/978-3-032-18582-2_9

of factor analysis, to derive scales to capture values of political and psychological kinds. Factor analysis is a statistical method used to describe variability among observed, correlated variables in terms of a potentially lower number of unobserved variables—called factors. The two dominant factors or scales that have historically been thought to capture political life are first, a scale of economic liberty or left versus right wing, and second, one of social liberty, variously labeled authoritarianism or communitarianism versus libertarianism or radicalism.

It is worth noting that some political scientists sideline political value spectra and have characterized a voting public using "identity-based" accounts, rather than "value-based accounts." These would rely more on inferring political positions based on demographic data about class, ethnicity, profession, and membership of certain community groups.

Model choice depends on its intended use and interpretation. It may be true that identity-based accounts better describe voting intentions and so have practical utility to a political strategist. But here, we are directly interested in people's values, and ultimately, in the real ethical questions of "how we ought to treat nature?" and to what extent the answers posed by philosophers to this question are palatable. While "values" have a more robust theoretical underpinning tracing a path from philosophy through to political science.

In our case, trying to infer value sentiments via an identity construct introduces conceptual muddiness. We could suppose there are identities in the vicinity of my subject—the "environmentalist," the "eco-warrior," the "anthropocentrist." But part of my argument in this book is to highlight stark divides *within* environmental thought, making such loose identities less tenable. To be an "environmentalist" may involve radical bioengineering à la David Pearce, or complete deindustrialization à la Arne Naess, or it may involve no positive sentiments toward the environment at all but rather a value for human welfare threatened by climate change. A model constructed around environmentalist identities could be predictively accurate, but it would not deepen understanding of people's thinking about the internal conflicts within environmentalism, and the differing solutions to climate change and biodiversity loss that these entail.

Meanwhile, as noted throughout this book, academia seldom focuses on technology as a source of intrinsic or political value. It is unsurprising then that we cannot find much empirical evidence on this point. We also noted that it is even more seldom the case that theoretical inquiry unites the areas of philosophy of technology and environmental ethics, so again, it is no surprise we find little empirical evidence uniting the two. The separation of enquiries gives no sight of possible correlations in attitudes between each domain of value in a single survey respondent, which is what would be required to demonstrate evidence of the theoretical scale we have built.

The only place we can find data on attitudes to both nature and technology is in the handful of large political attitude surveys, namely, the World Values Survey, European Values Study, European Social Survey, and Eurobarometer's main surveys. Unfortunately, these surveys strive for breadth over depth, and the few questions they contain on our subject matter do not provide the clarity we need. For instance, the fourth wave of the World Values Survey features the following questions:

> Now let's consider environmental problems in the world as a whole. Please, tell me how serious you consider each of the following to be for the world as a whole?
> [...] Global warming or the greenhouse Effect [....]
>
> Science and technology are making our lives healthier, easier, and more comfortable.
>
> Because of science and technology, there will be more opportunities for the next generation.
>
> All things considered, would you say that the world is better off, or worse off, because of science and technology?

Setting aside the ambiguity caused by the lumping together of science and technology, we can see that in all cases, the World Values Survey questions are only vaguely hitting upon values. Political values are the realm of the principled over the practical, the intrinsically valuable over the instrumentally valuable, distinguishing ends from means. These statements do not do this. To think global warming is serious, and technology comfort-inducing, opportunity-providing, and welfare-improving, only requires that I see how important nature and technology are as means to these various ends. But they do not necessitate a view on whether or not nature and technology are valuable as ends in themselves.

This inability to disaggregate the intrinsic from the instrumental value is a consistent theme across environmental value surveys such as the Global Environmental Survey, New Environmental Paradigm (NEP) index, and Schwarz's Environmental Values Survey. A review of environmental values scales by Bernstein and Szuster [1] noted that current metrics "*conflate a situated, incomplete set of environmental attitudes with pro-environmental worldviews generally*."

Another such survey, which seems to have been frequently published on by social psychologists Franz Bogner and Michael Wiseman, is the 2-MEV scale [3–7, 9–11]. I select 2-MEV rather arbitrarily as a subject for closer philosophical inspection, because it seems to be a popular and widely used scale, though I could have chosen numerous others which exhibit similar issues.

2-MEV posits there are two orthogonal dimensions of environmental value:

- Preservation—"*a biocentric dimension that reflects conservation and protection of the environment.*"
- Utilization—"*an anthropocentric dimension that reflects the utilization of natural resources.*"

The Preservation scale broadly captures sentiments and behaviors which indicate an intrinsic value for nature. The following scale items used in (2006) illustrate the point:

- *It is interesting to know what kinds of creatures live in ponds or rivers.*
- *It upsets me to see the countryside taken over by building sites.*

Meanwhile, the Utilization scale broadly captures people's attitudes to the use of nature as a means to other ends, such as material wellbeing.

- *Worrying about the environment often holds up development projects.*
- *Dirty smoke from chimneys makes me angry.*

- *Only plants and animals of economical importance need to be protected.*

These constructs of Preservation and Utilization are suggested (and evidenced) to be orthogonal. Multiple papers embrace this finding:

> The factors' orthogonality permits a respondent to vary his position on one dimension independently of that on the other. Hence the model implies no conflict between assigning high importance both to the protection of the environment and to the need to make use of natural resources—a configuration that demonstrably occurs in reality [8].

> It is essential to the 2-MEV that it does not imply a conflict between assigning importance to preserving the environment whilst recognizing the need to make use of natural resources (Bogner 2018).

Given the discussion of previous chapters, it strikes me as a conceptual and moral failure that these concepts are not negatively correlated, that Bogner and Wiseman do not see there being a trade-off between protecting and using the environment. Bogner and Wiseman make passing reference in a literature review to the philosophical origins of the eco-centric stance in Dewey, Muir, Leopold, and Carson (2018)). But it is a crucial battle-cry of these thinkers that nature ought not to be seen in exploitative terms. The extent to which we see nature as anything other than to be exploited ought to be the extent to which we want to preserve it.

Quite bizarrely, I can, on this scale, be a strong advocate of exploitation and of conservation. I can be a strong anthropocentrist and a strong biocentrist. This vastly reduces the scale's explanatory power: If, as a civil servant, I was tasked with estimating the public acceptability of building a new urban development project, I could evidence two contradictory conclusions on the basis of a 2-MEV study. I could observe that respondents are "*upset to see the countryside taken over by building sites,*" and conclude it is unacceptable. But I could also see the same respondents agree that "*worrying about the environment holds up development projects*" and conclude that I should continue without delay. Likewise, in deciding whether to invest public money into a nature conservation project, I might be buoyed by the finding that people are "*interested to know what kinds of creatures live in ponds or rivers,*" but puzzled that the same respondents say, "*Only plants and animals of economical importance need to be protected.*"

Let us consider some possible explanations for what has gone awry here.

9.2 Explanation 1: No Forced Choices

Suppose a town asked participants Q1: *Would you like more milk chocolate in the shops?* And Q2: *Would you like more breakfast cereals in the shops? Prima facie* these choices seem uncorrelated, and consistency demands nothing of one choice in relation to the other. Now suppose there is a severe milk shortage in the town and residents are aware. There are only a few goods containing milk left on the shelves. It now becomes more evident to the participant that in answering "yes" to Q1, they are implying they would like more of the rationed milk stock to be spent on producing chocolate at the expense of milk that they can buy to use on their cereal for breakfast. We would expect answers to no longer be uncorrelated but negatively correlated.

Now suppose again that the residents are not fully aware of the milk shortage, but democratic policymakers nevertheless need to gauge public preferences about the use/non-use of milk. The issue here is that people may well just opt for more of both items, which is unhelpful. The policymaker needs to force the survey respondent to make a choice, for instance, Q3: *would you prefer milk chocolate to breakfast cereal?*

Like the hypothetical town's milk supply, the real-world stock of nature is in short supply, presenting a problem for the policymaker. While someone may profess to like both using and preserving nature very much, the two goals are in conflict. But not everybody intuits this. Indeed, it appears that Bogner and Wisemann do not intuit this, as they not only endorse the orthogonality of use and preservation, but even go as far as to imply that the 2-MEV scales are parameters of environmental educational outcomes—i.e., a successfully educated individual *ought to* score highly on both preservation and utilization scales (e.g., 2006, 2018). Bogner and Wisemann do not include scale items explicitly forcing a trade-off, and so, there is no need for participants to ever express mutually consistent preferences.

The public's failure to grasp the consequences of environmental utilization is not surprising. It could be similar in structure to another phenomenon in the separate but related matter of solving climate change. Michael Hoexter is credited with formalizing the notion of "Soft Climate Denial." Some key thought patterns cited by Hoexter [13] contributing to soft climate denial include:

- *Compartmentalization*: everyday events lack an obvious connection to global warming, so it becomes an abstract item of knowledge not linked to practical action.
- *Climate Providentialism*: post-industrial societies are disconnected from nature, view it as an exogenous and eternally provider, regardless of drastic changes.
- *Millenarianism*: people become preoccupied with an eventual, large-scale transformation of society, not day-to-day concrete action.

Arguably, these same three thought patterns give rise to the positions expressed in these surveys that it is possible to both preserve and intrinsically value nature, and do the opposite. One might *compartmentalize* a love of the wild, and not see that it is, on a daily basis, threatened by those engaged in artifice. One might do so because the transition from one to the other takes place *out of view*. One might also have idealistic nature-loving tendencies and yet maintain, in another more practical, *day-to-day* frame of mind, that in fact they are grateful for the material comforts of modernity.

9.3 Explanation 2: Mismatch of Items and Theoretical Construct?

Another part of the explanation for the 2-MEV's mismatch between theoretical incompatibility and statistical orthogonality could be that one or other of the theoretical constructs does not actually describe the correlation captured in the scale items.

Let us first examine a list of Utilization scale items from [2] (items remain relatively unchanged since conception in the 1990s):

1. *We must build more roads so people can travel to the countryside.*
2. *Nature is always able to restore itself.*
3. *Our planet has unlimited resources.*
4. *We do not need to set aside areas to protect endangered species.*
5. *People worry too much about pollution.*
6. *The quiet nature outdoors makes me anxious.*
7. *We need to clear forests in order to grow crops.*

What could this theoretical construct be, if not as the authors argue, some notion of "Utilization"? One possible interpretation consistent with orthogonality is that items indicate a level of awareness or education about the processes involved in environmental economics. Questions 2, 3, and 7 seem to elicit respondents' knowledge of facts that indicate their understanding of environmental issues, such as that we sometimes do need to clear forests to grow crops. This awareness can feasibly be orthogonal to any evaluative stance on those facts, which would be covered in the Preservation scale. Another possible interpretation, particularly items 2, 3, 4, and 5, is that this scale measures one's optimism/pessimism about the planet's ability to cope with humanity's demands on it. Again, this scale would be orthogonal with the Preservation scale, as loving nature does not bear either way on the empirical matter of whether nature survives.

However, I think these items do seem to be best explained in the author's intended way: all items do seem united by an intention to exploit nature for some purpose or other. Now let us examine the Preservation scale.

Returning to the theme of instrumental versus intrinsic value, the 2-MEV Preservation scale is flawed in a similar vein to the WVS in focusing on nature's instrumental value in many of its questions. The following items again from [2] do all give some reason to preserve nature, but none necessarily entail eco-centrism:

- *I save water by taking a shower instead of a bath (in order to spare water).*
- *I always switch the light off when I don't need it.*

While indicative of someone being environmentally conscious, these behaviors also are consistent with simple economic prudence.

- *I enjoy trips to the countryside.*

There are many reasons why I might enjoy the countryside beyond having preservation value, for instance, because I can burn a log fire.

- *Dirty industrial smoke from chimneys makes me angry.*

Again, this could be because of the direct harms to humans caused by localized pollutants, rather than because of some intrinsic value for nature.

Secondly, some of the 2-MEV Preservation survey items survey the existence of factual beliefs which are logically independent of any evaluative stance:

- *Humankind will die out if we don't live in tune with nature.*
- So*ciety will continue to solve even the biggest environmental problems.*

Perhaps it might be argued that these empirical beliefs are commonly held by those who have an intrinsic value for Preservation, but then we would want further evidence linking factual beliefs to ideology, rather than presume a relationship between them. These factual claims might all be true, but that is irrelevant to the respondent's appraisal of the environment itself.

Thirdly, two scale items used in 2-MEV capture some notion of animal welfare:

- *Human beings are not more important than other creatures.*
- *Pets are part of the family.*

We saw in previous chapters that value for animals as moral subjects is conceptually distinct from intrinsic value for nature. The fact that we find empirical correlations between the two, as I argued, is an unfortunate and historically contingent confusion, and one that will not stand the test of time in an age of synthetic life. Animal rights enthusiasts have some common ground with environmentalists, but the gap is widening.

These issues with the items on the Preservation scale come together to make me think what is being measured here is not precisely eco-centrism, but rather, a cluster of attitudes and beliefs which happen to be held by those who would broadly associate themselves with an environmentalist movement. Recalling the discussion above of identity-based accounts versus values-based accounts, perhaps this scale is in fact measuring some loose "environmentalist identity," not environmental values as it purports to.

9.4 Explanation 3: One-Sidedness

Putting aside the issues of suspect constructs and orthogonality, my final issue with 2-MEV and other environmental values scales is a re-emphasis of the one I began this section with. All scales so far have been one-sided. They assume some purely positive construct, to describe the extent to which someone values the environment. The interpretation of a low score on these scales is unclear. It could be either complete ambivalence, or complete disagreement.

What I have argued is special about political values spectra is that they occur when there are genuinely incompatible "goods" that are located *at either end*. These cause disagreements upon which there can be passionate responses on both sides, rather than one side being ambivalent. There is a fundamental trade-off between economic liberty and economic equality, and between libertarianism and authoritarianism.

It could be argued that in some survey items, value for nature has been traded off against something incompatible: economic welfare. But we should see that economic welfare is not the direct opposite of "value for nature" in the way economic liberty is the opposite of economic equality. Economic welfare could be set against *any* intrinsic value. We could ask people whether they would be willing to pay much higher prices in the name of equality, or liberty, or community, or art, or honor, or honesty, or anything else. The parameter of economic welfare essentially describes our society's resources, and we face many options in how we employ those resources to various "ends." Economic welfare might be an end in itself, but it is also a very important, perhaps the most important, *means* to many other ends, and so it is valuable. But nature is just one such other end.

In this way, we could question whether "Utilization of nature" is actually a distinct political value at all? Or whether the value in question is just a value for a more general form of economic "utility," which could have, in another unrelated survey design, been contrasted with any other desiderata. I have advanced throughout this book that instead, the *intrinsic* value opposing nature, and so creating a political ideological spectrum, is an intrinsic value for technological endeavor, for taming the wild, for mastering nature's mysteries, and for the unparalleled creativity and productivity of mankind.

Moving on from Bonger and Wiseman's 2-MEV scale and other environmental value surveys, I would like to give one final clarifying note on the relationship between my spectrum and another popular political spectrum which may occur in the minds of political science readers, which is Ronald Inglehart's "Post-Materialism Spectrum."

9.5 Inglehart's Post-materialism

Political scientist Ronald Inglehart has found empirical evidence for a spectrum he calls materialism versus post-materialism. In The Silent Revolution (1977), he finds that the affluence of post-war generations in the west has led them to place greater emphasis on non-material goals such as autonomy, gender equality, freedom of speech, and, most importantly for our sake, environmentalism.

Prima facie, based on the words involved, we might consider whether the nature-technology spectrum could collapse into the materialism versus post-materialism spectrum, where "intrinsic value for technology" maps to materialism and "intrinsic value for nature" to post-materialism. However, this does not work. Inglehart's materialism does not mean "concerned with physical material" per se. It is not to be confused with the colloquial sense of "materialism" as in the dispositions of a vapid socialite. It is about meeting basic human needs like hunger and physical security. Notably, he considers "aesthetic satisfaction" a sub-category of *post-materialist* value even though we know such satisfaction will often be given by material objects, e.g., art. Meanwhile, my intrinsic value for technology is to be sharply distinguished from an instrumental value for technology. Whereas Inglehart's materialism sees its materials as instruments to satisfy other ends such as security and hunger.

The better way to reconcile the two spectra is to see Inglehart's as a sort of "meta-political spectrum," attempting to describe a movement from hard matters of human survival found in poorer countries and times, to a more abstract value-laden politics in richer countries and times, who possess the luxuries of leisure time, education, and democracy to engage in a more philosophical political discourse. Within societies Inglehart would characterize as highly post-materialist, such as the UK, there are ideological disagreements. These are internal to post-materialism. My nature-technology spectrum is, then, one such internal post-materialist disagreement. So, when Inglehart finds evidence that post-materialist societies are more likely to take up the cause of environmentalism (which we can suppose might overlap with my "intrinsic value for nature" parameter), I would imagine that the same post-materialists would also be more likely to take up the rival cause of "intrinsic value for technology."

9.6 A Suggestion for Future Empirical Research

Here we demonstrate that existing political science scales are fundamentally unfit to capture the nature-technology political spectrum. Their reliance on instrumental and welfare-based measures fails to isolate the core axiological conflict necessary to test the theoretical claims of this book. To correct this methodological shortfall and empirically validate the nature-technology spectrum's existence as a third, independent dimension of political values, a dedicated research program is required.

A new survey scale must address the flaws identified in the 2-MEV and other existing environmental scales. Scale items must exclusively target the intrinsic value of the two poles. Questions must avoid conflating technological progress with instrumental benefits (e.g., job creation, economic comfort) or natural preservation with instrumental human welfare (e.g., personal safety from pollution).

Second, the design must move beyond simple ambivalence. Questions must require respondents to choose between two mutually exclusive goods located on opposite ends of the spectrum, thereby forcing them to express their dispositional preferences. For example, framing a trade-off between pristine wilderness (naturalness) and the construction of a complex scientific facility (technology) in the same location.

Third, items must explicitly focus on the themes established in this work. Nature questions should focus on authenticity, giftedness, awe, preservation, and non-interference with natural systems. While technology questions should focus on mastery, ingenuity, conquest of limits, the value of the act of creation, and collective human achievement.

To validate or invalidate the theoretical claims of this book, the subsequent statistical analysis must reveal the orthogonality of the nature/technology items with established Economic Left/Right and Social Libertarian/Authoritarian factors. The nature and technology scale items must exhibit high internal consistency, confirming the coherence of the two opposing value clusters. I encourage political scientists to take up the mantle.

References

1. J. Bernstein, B. Szuster, The new environmental paradigm scale: reassessing the operationalization of contemporary environmentalism. J. Environ. Educ. **50**(2), 73–83 (2019). https://doi.org/10.1080/00958964.2018.1512946
2. F. Bogner, Environmental values (2-MEV) and appreciation of nature. Sustainability **10**, 350 (2018). https://doi.org/10.3390/su10020350
3. F.X. Bogner, M.G. Wilhelm, Environmental perception of pupils: development of an attitude and behavior scale. Environmentalist **16**, 95–110 (1996)
4. F.X. Bogner, M. Wiseman, Environmental perception of Danish and Bavarian pupils: towards a methodological framework. Scand. J. Educ. Res. **41**, 53–71 (1996)
5. F.X. Bogner, M. Wiseman, Environmental perception of rural and urban pupils. J. Environ. Psychol. **17**, 111–122 (1997)
6. F.X. Bogner, M. Wiseman, Environmental perception of Swiss and Bavarian pupils. Swiss J. Sociol. **24**, 547–566 (1998)
7. F.X. Bogner, M. Wiseman, Towards measuring adolescent environmental perception. Eur. Psychol. **4**, 139–151 (1999)
8. F.X. Bogner, M. Wiseman, Environmental perception: factor profiles of extreme groups. Eur. Psychol. **7**, 225–237 (2002)
9. F.X. Bogner, M. Wiseman, Environmental perception of pupils from France and four European regions. J. Psychol. Educ. **1**(7 & 1), 3–18 (2002)
10. F.X. Bogner, M. Wiseman, Outdoor ecology education and pupils environmental perception in preservation and utilization. Sci. Educ. Int. **15**(1), 27–48 (2004)
11. F.X. Bogner, M. Wiseman, Adolescents attitudes towards nature and environment: quantifying the 2-MEV model. Environmentalist **26**, 247–254 (2006)
12. H.J. Eysenck, S. Eysenck, *Manual of the Eysenck Personality Questionnaire*. (Hodder and Stoughton, 1975).
13. M. Hoexter, *A Pocket Handbook of Soft Climate Denial*. (New Economic Perspectives, 2016). http://neweconomicperspectives.org/2016/10/pocket-handbook-soft-climate-denial.html
14. R. Inglehart, *The Silent Revolution: Changing Values and Political Styles Among Western Publics*. (Princeton University Press, 1977)
15. F.N. Kerlinger, The structure and content of social attitude referents: a preliminary study. Educ. Psychol. Measur. **32**(3), 613–630 (1972). https://doi.org/10.1177/001316447203200305
16. C.P. Middendorp, *Progressiveness and Conservatism: The Fundamental Dimensions* (1978)
17. M. Rokeach, *The Nature of Human Values* (Free Press, 1973)
18. B. Todosijevic, Structure of political attitudes a literature review, in *Political Attitudes and Mentalities. Eastern European Political Cultures: Modeling Studies* ed. by C. V. Voinea, B. Todosijevíc, G. Boella. (ArsDocendi-Bucharest University Press, 2013), pp. 23–52

Chapter 10
Conclusion

Abstract This book concludes by consolidating the central philosophical claim: that a wide array of esoteric philosophical, ethical, and public policy debates at the intersection of nature and technology are best understood through the simplifying framework of a Nature-Technology political spectrum. The chapter reviews the framework's foundation: the conceptual apparatus of a political spectrum, which holds that political values are intrinsic, non-instrumental goods that exist in antagonistic pairs (Nature as "absence of human involvement" versus Technology as "intentional functional artifact"). It summarizes the original arguments built for both poles, particularly distinguishing the intrinsic value of naturalness from the moral consideration of non-human personhood, and defending technology's value through appeals to physical liberty and technological achievement. The conclusion confirms that the spectrum successfully met all four blind judging criteria: demonstrating genuine metaphysical conflict, robust explanatory power across environmental and bioethical policy (e.g., human enhancement), validity as a continuum (not a binary), and timelessness. The utility of the political lens is reaffirmed as a tool to navigate the "messy middle" and quash the demand for "lines in the sand" in debates like human enhancement. Finally, a speculative vision of a post-human future is offered to underscore the spectrum's enduring relevance as a structural feature of political disagreement that is projected to outlive humanity as we know it.

Keywords Political spectrum · Intrinsic value conflict · Nature-technology trade-off · Human enhancement debate · Political axiology · Future of politics · Post-humanism

The main philosophical claim of this thesis is that many *prima facie* esoteric philosophical, ethical, and policy debates in the once-disparate vicinities of nature and technology can be better understood with reference to a simplifying framework: a spectrum of intrinsic value covering, on the one hand, value for nature and on the other, a value for technology.

This book has attempted to build up a framework for understanding and arguing about a wide range of political issues, spanning lots of sub-fields of philosophy, while

B. Steyn, *Technology vs Nature*,
SpringerBriefs in Applied Sciences and Technology,
https://doi.org/10.1007/978-3-032-18582-2_10

also remaining focused on the key claim. We began by laying down a conceptual apparatus: a political spectrum—a device severely underused by political philosophers. We saw that political values are intrinsic values—things valued for their own sake. And political values are distinctive among intrinsic values because they relate to one another in antagonistic pairs. A spectrum helps create understanding in debate, something that those on, for instance, the left and right take for granted when arguing with the other. This understanding of the precise relationship between different political ideologies is often lost in the transition from political science to political philosophy.

We then moved to explain our desiderata for this particular political spectrum: nature and technology. Nature, the absence of human involvement. Technology, intentional functional artifacts, where artifacts are things which have been made with human involvement. It is clear from just stating these definitions how the two are in tension with one another.

We reviewed the field of environmental ethics and how it has argued for nature's intrinsic value. We made a crucial distinction here: nature's value is separate to the value placed on non-human animals. It is the naturalness we are interested in here.

Next, we build an original case for technology's intrinsic value. First, appealing to real-world intuitive examples like spacecraft. Next, we saw three theoretical arguments: The first was about how technology offers a unique form of liberty, which is analogously valuable to how eminent philosophers like Amartya Sen have argued for other forms of liberty (social, economic) being valuable. The second was an argument that the act of technology production is a virtue, like how Virtue Epistemology theorists like Ernst Sosa have thought about knowledge. The third looked at how technology's intuitive value has often been misattributed as a value for scientific knowledge.

Next, we saw the spectrum in action with a deep dive into the lively world of human enhancement. My claim is simply that the debate over human enhancement's unnaturalness is just one instance of the wider values trade-off we have outlined in previous chapters, concerning the way humans interact with not just other humans, but all physical material.

We then zoomed out to view the current theoretical landscape discussing nature and technology. We see that political science has so far not been able to properly isolate the political value for nature and technology, because it conflates matters of value with more practical and economic considerations. Meanwhile, neither analytic nor continental philosophy have yielded the precision and unifying aspects of this framework. We saw the failure of current empirical work in isolating this phenomenon. This shows us that the present work is, quite surprisingly given its simplicity, original in its claim.

At the start of this book, I suggested four blind judging criteria with which to judge a candidate political values spectrum. The first test was whether both sides of the proposed spectrum are in genuine conflict. We saw how the metaphysics of nature and technology deeply oppose. Even if, for whatever reason, the metaphysical argument was not accepted, we also saw that empirically, scholars and members of the public do adopt and debate these conflicting positions across a range of policy debates.

The second test was about explanatory power. The new spectrum should explain a significant portion of the variance in attitudes to a significant number of prima facie separate political issues that cannot be explained by incumbent political spectra. Across this book, we've seen how the spectrum can be deployed to diagnose debates in environmental policy, bioethics, industrialization, and technology policy. We saw in a comparison of utopias that the nature-technology conflict does not reduce to existing dimensions of political disagreement, e.g., left versus right. We also saw that technological progress is of-a-piece with the social and economic realms for offering a plane of liberty, which is encouraging, but also that the plane of liberty is distinct from the former two.

The third test was whether the matters and values in question can be rightly rendered on a spectrum, not a binary. We see the spectromic perspective here, because the framework we have is built on an underlying metaphysical continuum where nothing is truly natural or truly artificial. Extreme examples on either side should be enough to make most readers intuit they are somewhere in the middle terrain of this spectrum, whether it be the misanthropic consequences of total eco-centrism, on one hand, or the soulless destruction of everything we would want to call human nature on the other. We saw that a spectromic perspective is particularly useful (and necessary) in human enhancement debates because it wards off slippery slope arguments and quashes bioethicists' absurd demand to draw a "line in Darwin's sand"—i.e., define some magical point at which human enhancement suddenly becomes wrong.

The final test was timelessness. We have here a tension once thought to be unique to the 18th Century's Enlightenment and Romanticism, which I have severed from those historical roots. This timelessness is perhaps the most salient feature of all. We see that while the policy specifics may change, fundamental debates around economic and social liberty run back as far as politics itself. The nature-technology political spectrum may be late to the party, but it looks set to stay for a long time, outliving humanity as we know it.

Imagine the year is 2100. Genetic modifications have swept across the world. A procedure is routinely performed on babies to increase their base levels of serotonin. Most people regulate their mood with drugs and artificial pleasures. The natural world is an anachronism. Non-human biology is a canvass for technological creativity.

There is some resistance to this trend. Some get a palpable itch that they are living in designed, ersatz realities. They yearn for something authentic and real. They yearn for a natural experience. Meanwhile, others see the technological turn as our true calling. Technological progress, having performed miracles on poverty, disease, hunger, and other 21st-century problems, becomes a quasi-religious mentality. With some such apostles editing and manipulating themselves beyond recognition, the human category is breached. It becomes unclear who politics is for.

Those striving to protect natural experience will accordingly strive to deny these post-humans rights and recognition. But the pro-technologists and indeed post-humans themselves will disagree. Political suppression of post-humans leads to violence and war. With one side embracing enhancement, and the other taking a principled opposition, there could only ever be one winner.

Only 30 years later, a post-human discovers a way to upload brains to the cloud. Schisms split between those who are literally and figuratively attached to their bodies as a remnant of a primate origin, and those who are keen to embrace the new way. The debate rages on…

MIX
Papier aus verantwortungsvollen Quellen
Paper from responsible sources
FSC® C105338

If you have any concerns about our products,
you can contact us on
ProductSafety@springernature.com

In case Publisher is established outside the EU,
the EU authorized representative is:
Springer Nature Customer Service Center GmbH
Europaplatz 3, 69115 Heidelberg, Germany

Printed by Libri Plureos GmbH
in Hamburg, Germany